Nikolaos Gavanas

Towards an integrated accessibility model for the Balkan region

AF138222

Nikolaos Gavanas

Towards an integrated accessibility model for the Balkan region

Concepts and methodological framework

LAP LAMBERT Academic Publishing

Impressum / Imprint

Bibliografische Information der Deutschen Nationalbibliothek: Die Deutsche Nationalbibliothek verzeichnet diese Publikation in der Deutschen Nationalbibliografie; detaillierte bibliografische Daten sind im Internet über http://dnb.d-nb.de abrufbar.
Alle in diesem Buch genannten Marken und Produktnamen unterliegen warenzeichen-, marken- oder patentrechtlichem Schutz bzw. sind Warenzeichen oder eingetragene Warenzeichen der jeweiligen Inhaber. Die Wiedergabe von Marken, Produktnamen, Gebrauchsnamen, Handelsnamen, Warenbezeichnungen u.s.w. in diesem Werk berechtigt auch ohne besondere Kennzeichnung nicht zu der Annahme, dass solche Namen im Sinne der Warenzeichen- und Markenschutzgesetzgebung als frei zu betrachten wären und daher von jedermann benutzt werden dürften.

Bibliographic information published by the Deutsche Nationalbibliothek: The Deutsche Nationalbibliothek lists this publication in the Deutsche Nationalbibliografie; detailed bibliographic data are available in the Internet at http://dnb.d-nb.de.
Any brand names and product names mentioned in this book are subject to trademark, brand or patent protection and are trademarks or registered trademarks of their respective holders. The use of brand names, product names, common names, trade names, product descriptions etc. even without a particular marking in this works is in no way to be construed to mean that such names may be regarded as unrestricted in respect of trademark and brand protection legislation and could thus be used by anyone.

Coverbild / Cover image: www.ingimage.com

Verlag / Publisher:
LAP LAMBERT Academic Publishing
ist ein Imprint der / is a trademark of
OmniScriptum GmbH & Co. KG
Heinrich-Böcking-Str. 6-8, 66121 Saarbrücken, Deutschland / Germany
Email: info@lap-publishing.com

Herstellung: siehe letzte Seite /
Printed at: see last page
ISBN: 978-3-659-52344-1

Zugl. / Approved by: Thessaloniki, Aristotle University of Thessaloniki, Diss., 2011

Acknowledgements

I would like to express my gratitude to Professor Magda Pitsiava-Latinopoulou, Director of the Transport Engineering Laboratory of the Aristotle University of Thessaloniki, for the valuable mentoring and constant support in the research activities that led to the conduction of this manuscript.

Table of contents

List of tables

List of figures

List of equations

1. INTRODUCTION

The concept of a region in spatial analysis does not coincide with an administrative region but refers to a geographic entity of common or complementary territorial and socio-economic features (OECD, 2002). A region includes a system of activity poles, such as urban centres, industrial areas and transport gateways, which interact with each other and the rest of the world on the basis of administration and services, production and commerce, labour market and natural resources, social interaction and cultural heritage. In this context, the role of the regional transportation system is to provide the appropriate accessibility conditions for the service of the flows of people and goods within and throughout the region.

The Balkan region, which is the study area of the current research, includes the countries Slovenia, Hungary, Romania, Croatia, Bosnia and Herzegovina, Montenegro, Serbia, Bulgaria, Albania, F.Y.R.O.M and Greece. A common historical fact for the region is that it comprised the European territory of the Ottoman Empire from the 16th to the 19th century. The foreign occupation has led to the lagging of socio-economic development in the Balkans while at the same period significant changes were taking place in Western Europe, preparing the ground for the industrialisation era and setting the basis for playing a key-role in the contemporary globalised market. The end of the occupation was followed by a period of warfare among the Balkan nations creating the conditions for further divergence. The gap between the Balkan region and the rest of Europe was increased after the 2nd World War due to the integration of the Balkan countries to the system of planned economies of Eastern Europe, with the exception of Greece which became a NATO ally. After the end of the Cold War and during the 90's, almost the whole of the Balkan region was rumbled by a new period of socio-political turbulence and warfare. At the meantime, Greece was the first Balkan country that begun the transition of becoming part of the common European territory. In 2003, more than a decade later, the 5th enlargement of the European Union (EU) was initiated with the gradual accession of the rest of the Balkan countries (Getimis & Kafkalas, 2007). A main challenge of the enlargement referred to the achievement of territorial cohesion throughout the European Union and the socio-economic convergence of the new member states. Towards this purpose, a significant effort is still being conducted through the cooperation of the European Union and the Balkan countries for the upgrade of transport infrastructure, the reform of the transport policy and the restoration of the accessibility conditions of the Balkan region. Five years after the enlargement, the globalisation of the financial crisis sets new obstacles in the course of convergence between the European regions and compromises the international competitiveness of the enlarged European Union.

In the midst of the on-going changes at the regional, the European and the global level, the scope of the current research is the presentation of a methodology for the development of an integrated potential accessibility model for the transportation system of the Balkan region. The following objectives are set:

- The investigation of the role of transport accessibility in promoting regional development,
- The discussion of the significance of the assessment and monitoring of accessibility in the Balkans and
- The description of the appropriate framework of components for the formulation of a potential accessibility model for the Balkan region.

The structure of the research is presented in Figure 1. Each step of the research is used as an input for the following step and corresponds to each Chapter of the manuscript. According to the Figure, a description of the scope of the research and the structure of the methodological approach is given in the introductory part, followed by a brief review of the development of the transportation system from the industrialisation to the globalisation era as a contributor towards regional development. Next, the concept of regional accessibility is described in relation to its effect on peripherality, providing the theoretical background for the policy analysis of the next step. In specific, the step aims at the analysis of the European Union's policy framework in the sectors of enlargement, convergence, territorial cohesion and transport, which are directly related to the strategies for the formulation of the accessibility conditions in Europe. The step is concluded with the synthetic analysis of the policy framework in order to allocate the priorities for the transportation system of the examined region. The aforementioned priorities and the features assessed by the use of the appropriate indicators are used for the evaluation of the transportation system of the Balkans by the implementation of a S.W.O.T analysis. The synthetic results from the

S.W.O.T. analysis are used in order to define the components of the model for the assessment of potential accessibility in the Balkan region. Apart from the specific features of the region, the proposed components are also based on the state of the art from the development of accessibility models in Europe and the conduction of a preliminary case study for the investigation of the changeability in the accessibility conditions of the Balkans due to accession process and the gradual completion of the major transport infrastructure projects. The conclusive step of the research refers to the discussion of the main findings, the necessary features for the development and implementation of the proposed potential accessibility model and the prospect of the model as a tool for the support of decision making and strategic planning of the Balkan transportation system.

The research is based on the author's PhD thesis entitled: "Spatial impacts of the transport system: Implementation for the wider area of South-East Europe", which was concluded in 2011 under the supervision of Professor Magda Pitsiava-Latinopoulou for the Aristotle University of Thessaloniki, Greece (Gavanas, 2011). The thesis focuses on the assessment and evaluation of the transportation system of Southeast Europe and the related spatial impacts during the EU enlargement era, by developing and implementing a system of 54 socio-economic and transport indicators and a potential accessibility model, which is presented in the preliminary case study of the current manuscript. Furthermore, a series of non-funded research has been conducted in the context of scientific publications in journals and conference proceedings for the dissemination and update of the PhD findings. These publications also provide significant input for the research. Finally, the research capitalises from the experience in the research projects: "Spatial impacts of multimodal corridor development in gateway areas: Italy-Greece-Turkey, (SIMCODE-IGT)" and "Southeastern Mediterranean spatial observatory network, (SEMSON)" in the framework of Archimed 2000-2006, Interreg IIIB.

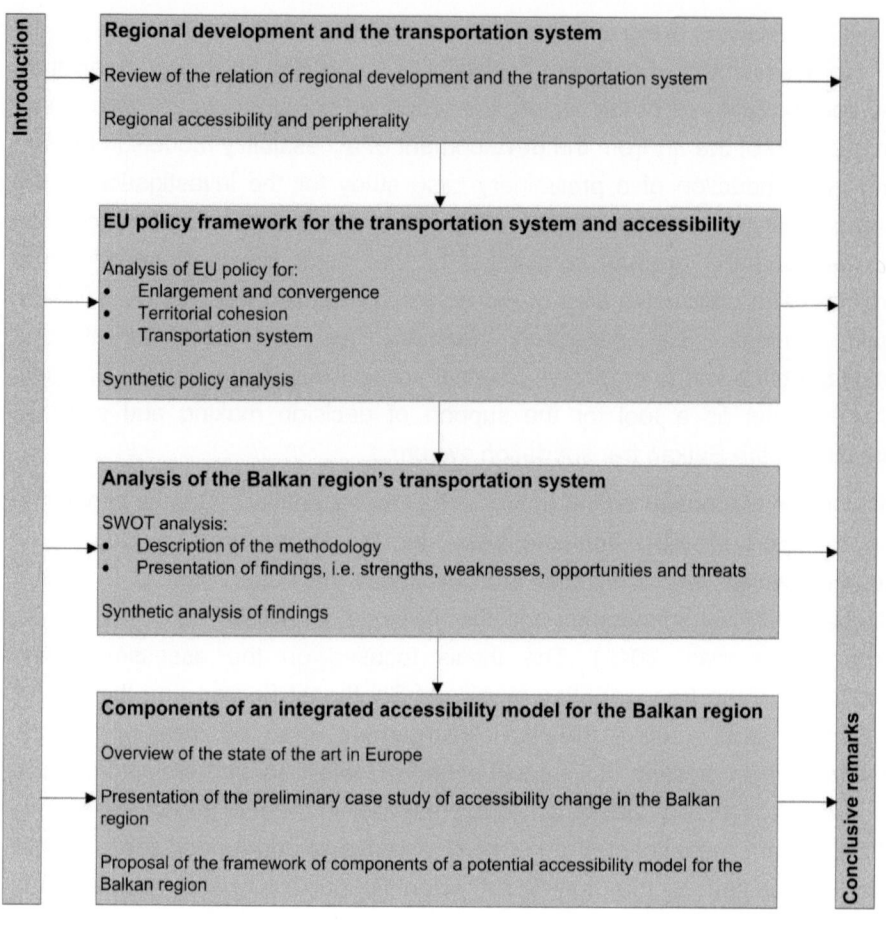

Figure 1. Structure of the research

Source: Own elaboration

2. REGIONAL DEVELOPMENT AND THE TRANSPORTATION SYSTEM

A specific socio-economic activity or group of activities, such as work, education, recreation etc., is conducted in a specific location, i.e. an activity pole. The spatial distribution of the activity poles and the relation between them formulate the network of activities at the local, regional and international level. The network of activities changes with the process of socio-economic development. The time and spatial separation of the activity poles generates the mobility needs of people and goods. The main driving force for the development of the transportation system is the need to overcome this separation and provide access to the activity network according to the conditions and requirements of socio-economic development (Krugman, 1998). On the other hand, the development of the transportation system creates new opportunities for socio-economic development due to the increase of mobility and the ability to access new activities. Consequently, the features of the transportation system are formed through time according to the socio-economic needs and expectations and vice versa.

The current chapter discusses the relation between regional transportation and socio-economic development with the aim to establish a theoretical background according to the scope of the research. It focuses on the notion of transport accessibility and its impact on regional development. More specifically, the Chapter includes the brief review of the development of the regional transportation system in relation to the socio-economic conditions and the conceptual clarification of transport accessibility in relation to regional development and peripherality.

2.1 Review of the relation of regional development and the transportation system

2.1.1 The transition from the industrialisation to the globalisation era

In each period of the evolution of the transportation system, specific transport modes are favoured by the socio-economic conditions. These modes develop either competitively or complementarily in relation to the rest of the transportation system. For example, the development of the Western European colonies from the 16th to the 18th century was based on the transportation of people and goods overseas with the disadvantage of lacking effective land transport links. During the industrial revolution of the 19th century, the inland waterways and the railways were linked to the seaport network and provided access to the mainland. Thus, specific regions of Western Europe and the New World gradually composed a more flexible network of socio-economic interactions and started to play a key-role at the international level (Rodrigue, 2013).

Almost a hundred and fifty years later, the post-World War II era introduced the dominance of road transportation and the rapid development of the motorway networks, which were connected to the seaports, the industrial areas and the urban network and substituted at a large degree the railways in the provision of intra and inter regional access (Baldwin & Baldwin, 2004). In many economically developed regions, such as these of Western Europe and even more the United States, the competitive advantages of the automobile system in terms of flexibility and door-to-door service diminished the role of other modes in short and medium distance passenger and freight transportation. During the globalisation of the economy, these advantages strengthened the key-role that these regions had acquired during the industrial period, while they increased the competitiveness of the emerging economies. An example of the role of road transportation in the regional development of westward societies during the last decades can be found in the ESDP (European Spatial Development Perspective), where it is mentioned that in the late 20th century a large-scale shift was observed in the Central and Eastern European countries from eastward to westward due to the process of EU accession, which was expressed by the shift from the rail to the road transportation system and from public to private transport (Committee on Spatial Development of the European Commission, 1999).

Since the 1980s, the needs of the globalised economy led to the development of the contemporary supply chain in freight transportation. Innovative transportation logistics combined with the liberisation of the transport market allowed the intermodal use of short and long distance transport modes and the integration of the supply chain from the phase of production to the phase of local distribution to the end-consumers (Alfalla-Luque & Medina-Lopez, 2009).

The end of the 2nd World War and the beginning of the globalisation era also initiated the intense development of another dominant transport mode dedicated to long distance passenger transport, i.e. the air transportation system. The competitiveness of the air transport market, the constant upgrade of the aviation technology in terms of infrastructure and services and the expansion of the international airport network with nodes usually located at the outskirts of main urban centres presented new opportunities for regional development (Wei & Yanzi, 2006). Apart from the air transportation system, another mode for long distance passenger trips is the high speed railway system. Nowadays, there is an on-going effort for the wide implementation of high speed railways, which is restricted by the high investment cost and the landscape of each region (Albalete & Bel, 2010; Levinson, et al., 1999).

At the meantime, the emergence of transnational cooperation and agreements created the institutional and legislative framework as well as the infrastructure for the improvement of access to the international market and the poles of socio-economic activity in general. The development of inter-regional transport infrastructure became a priority for international funding bodies, such as the World Bank and the European Investment Bank.

In the discussion on the relation between the transportation system and regional development, a specific topic that should be addressed is the role of urban development and the impact of cities, which accommodate the majority of regional population and activity. In the EU regions for example, the 60% of the population and 85% of the European GDP refer to urban centres (European Commission, 2007a). Until the industrialisation period, the urban areas were relatively isolated walking cities of mixed land-use. In less than a century, the continuous flow of population towards the cities, where the industrial and commercial activities were concentrated, resulted at the enhancement of their size and power while increasing the demand for more efficient mobility services. At the first half of the 20th century, the massive transportation system and, then, the private car changed the concept of urban travel forever[1]. During this period, mechanised transportation affected the form of urban development due to the increased capacity and efficiency and the ability to service more complex mobility needs (Newman and Kenworthy,

---

[1] The promotion of car ownership was a main difference in the travel culture between Eastern Europe (and part of the Balkans) and Western Europe (i.e. the pre-enlargement European Union). Until the late 20th century and for almost fifty years, the structure of urban and regional development in the two sides of the Iron Curtain was influenced by this difference.

1999). Furthermore, large cities with dense transport networks became the centres of regional development while the metropolitan areas surpassed their regional significance and turned into landmarks of the globalised economy (Maki, 2002). At the same time, a series of negative impacts from this form of development were observed, i.e. urban sprawl, congestion, spatial segregation, social exclusion and degradation of the urban environment. In most cases, the severity of these impacts was proportional to the development rate of the city. During the last three decades, the problem of urban congestion has become one of the major drawbacks for regional development. In order to cope with the problem, there is an on-going effort of transport experts and stakeholders for the development of effective policies and interventions and their integration from the international to the local level, as it is further discussed in the following Section.

2.1.2 The contemporary challenges of the regional transportation system

Three main challenges can be identified for relation of the transportation system with regional development from the late 20th century until today. The first challenge is related to the high rate of socio-economic development, the expansion of the supply and demand features of motorised (mainly road and air) transportation and the severe or even irreversible external impacts which are imposed on the natural and built environment. As far as transportation is concerned, these impacts can be observed locally, such as pollutant emissions in congested transport nodes and along road segments, polluted coastlines due to the freight handling of seaports, extreme noise and emission levels from the operation of airports, segregation of natural sites due to the development of land transport axes etc. However, the overall effect of the transportation system on the environment is global as the local problems affect the climate change, the biodiversity, the energy consumption and dependency and the land take of transport infrastructure (Nemry & Demirel, 2012; Geneletti, 2003; Brown, 2001).

The second challenge refers to the technological innovation which comprises a competitive market with its own progressing mobility needs, while it influences decisively the expectations of people regarding the level of mobility and the ability to access new products and destinations. Moreover, the constant improvement of vehicle engine technologies as well as the implementation of new technologies in the fields of the fleet and traffic management and the quality of infrastructure and services improves the transportation system in terms of cost, efficiency and external impacts.

The third challenge concerns the intensification of the globalised economy which continues to update the goals of the transportation system for accessing new destinations and resources and for enhancing the competitiveness of the transportation market. Nowadays, the global economic crisis is shifting the balance in the economic map with the degradation of some previously developed economies, the emergence of new regions as key-players in the international competition and the lagging of progress in the developing regions (United Nations, 2009).

In the first decade of the 21st century, the notion of sustainable development, defined by the European Commission (2001a) as the development that meets the needs of the present without compromising the ability of future generations to meet their own needs, was adopted as the goal for transport development. The goals of sustainability, i.e. economic competitiveness, social welfare and environmental safeguard, are answering directly to the above challenges of the contemporary transportation system.

In this context, the regional transport policy and strategic planning focuses on the effort of restoring the missing links between regions for the improvement of cohesion, the enhancement of competitiveness and the promotion of intermodality with the scope of achieving the balanced development of the transportation system and the minimisation of its external costs. Furthermore, a main objective is the promotion of innovation and new technologies for the continuous upgrade of transport competitiveness (Nagurney, 2000; Finel & Tapio, 2012). A more sustainable approach is also adopted gradually for urban transportation, aiming at an innovative and efficient multimodal system where active and massive transportation will regain their role over the private car dominance.

Since 2008, the crisis is continuing to generate worldwide impacts on every aspect of the regional and inter-regional transportation system, i.e. the decrease of demand as a consequence of the lower income level and the drop of the international trade, the difficulty in the allocation of funds to support investments for the development, upgrade and management of infrastructure and services and the lower rate of competitiveness within the transportation market (Macario & Van de Voorde, 2009). At the urban level, the decrease in socio-economic activity seems to facilitate the modal shift from the private car to cheaper transportation modes, i.e. public and active transport. Nonetheless, this is not due to the increase of society's awareness regarding the advantages of sustainable mobility rather than to the reduction of the individual's economic capability. In addition, in many developing regions, local authorities are unable to fund projects that would upgrade their transport networks (ICLEI – Local Governments for Sustainability, 2011). The threats imposed by the global economic crisis are today's most urgent priority and stakeholders commit a continuing effort worldwide to appropriately assess and analyse the problem and formulate feasible solutions (International Transport Forum, 2009; European Commission, 2011a).

An attempt to summarise the main parameters that compose the framework for the development of the contemporary regional transportation system is presented in Table 1.

Table 1. Parameters of the development of the transportation system

Features of the transportation system		Type of transport	
		Passenger	*Freight*
Mode choice	*Short distance*	Private motorised	Truck
		Public	Rail
		Private non-motorised	Inland waterways
		Combined	Combined
	Long distance	Air	Ship
		High speed rail	Air*
Development of infrastructure		In relation to the urban development	In relation to the supply chain
Strategic development priorities		Competitiveness	
		Innovation	
		Sustainability	
		Global crisis	

* Used for the transportation of specific type of goods, such as high value goods, mail etc.

Source: Own elaboration

2.2 Regional accessibility and peripherality

In the globalised economy, the potential for economic growth leads to the expansion of transport links between developed and developing regions in order to access new markets and resources and formulate a network of international competitiveness. As a result of economic development, the higher level of social welfare leads to the demand for a larger volume and variety of goods as well as the communication of new ideas in terms of science and technology, cultural and social movements etc. The transportation system contributes to the networking of the regions and the access to the poles of socio-economic activity in order to fulfil the economic needs and societal expectations (Dargay, et al. 2007). Moreover, the use of transport modes itself turns into an indicator of the level of regional development, such as the increase in private car use in developing regions or the high share of active transportation in many developed regions.

Despite the fact that there are common socio-economic trends which can be identified in the development of the regional transportation system worldwide, it should be highlighted that this development is not balanced. The economically developed regions have a dense multimodal network that link their intra-regional activity poles to inter-regional destinations. The main problems in these regions stem from congestion and refer to the decrease of the transport efficiency and the related external impacts on the society and the environment (Frank, 2000). On the other hand, underdeveloped regions may suffer from similar local bottlenecks but their most significant deficiency is the inadequate intra-regional density of the transport network and the lack of sufficient international access (Jacobsen & Kristiansen, 1992).

This differentiation is observed in Europe and comprises a significant handicap for the balanced development of the transportation system. In order to illustrate this argument, in Figure 2 there is a presentation of two maps, a map of the GDP (Gross Domestic Product) per capita per country, which functions here as an indicator of economic growth, versus a map of the European multimodal transport network. A first conclusion from the comparison of the two maps is that the shape of the European transport network is able to outline the geographical boundaries of Europe. This means that there are corridors and nodes that connect the mainland with the remote outskirts of the continent. A closer observation of the spatial distribution of the transport network reveals that its density decreases gradually from the Northwest and West-central regions towards the Balkan and Mediterranean regions. A similar distribution is observed for the GDP per capita per country[2].

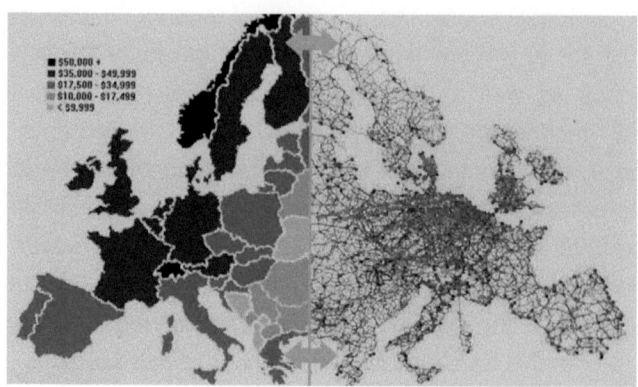

Source: Own elaboration; Maps: Zhang, et al., 2013 & http://commons.wikimedia.org/wiki/File: Europe-GDP-PPP-per-capita-map-worldbank.png accessed 30/10/2013

Figure 2. European GDP per capita vs European multimodal transport network

In order to analyse the differentiation in the availability of transport infrastructure between regions of different economic potential and its impact on the perspectives of regional development, the concepts of accessibility and peripherality should be clarified. Several definitions are given to the term "accessibility" in scientific and policy documentation. In an effort to synthesize some of these definitions at the regional level, it can be defined that accessibility is a result from the development and operation of the transportation system, which expresses the ability of a region to integrate its intra-regional activities and attract or approach the activities conducted in other regions (Littman, 2003; Spiekermann & Neubauer, 2002; OECD, 2002). Thus, a high level of accessibility is considered a location advantage of a region in relation to the other regions (Spiekermann & Wegener, 2006a).

---

[2] In the Scandinavian peninsula, the lower density of corridors versus the high rate of GDP can be attributed to the lower density of population and the landscape barriers (European Environmental Agency, 2011).

In this way, accessibility comprises the common ground of interaction between the spatial organisation of activities and the operation of the transportation system from the local to the international level (Figure 3). In Figure 3, the local level refers to the urban centres or other activity poles, where transport accessibility expresses the ability of a person to access the land uses of daily activity, e.g. work, education, recreation, health etc., and also the ability to distribute goods locally. At this level, efficient accessibility has a direct impact both on the local economy and the way of life. At the next level, a transport corridor interconnects a set of activity poles, settlements and resources and composes an intra-regional development corridor. At the regional level, the transport network with its corridors and nodes integrates the network of activity poles, settlements and resources into a unique spatial entity, i.e. the region. At this point, accessibility contributes to the cohesion and balanced development within the region. At the final level, the regional transport network is linked to the international transport network providing access to other regions. The sufficient accessibility of the region provides opportunities for international competitiveness and convergence. All levels of accessibility are related and play a significant part in regional development.

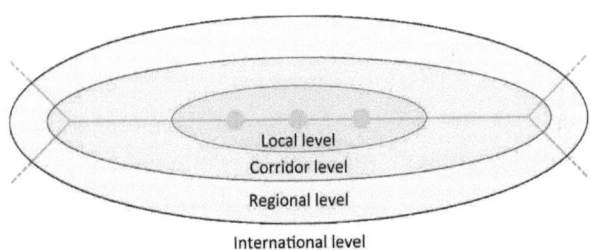

Source: ARE Transport and Infrastructure Planning and Transport Policy Sections, 2003

Figure 3. Levels of transport accessibility

On the other hand, the scarcity of infrastructure or the inefficient operation of the transportation system may lead to the increase of peripherality in terms of the separation of the activities within the region and the isolation of the region from the international competition (Schurmann & Talaat, 2000). Thus, peripherality can be expressed as the reverse function of accessibility (Panebianco, 2001). Until recently, the peripheral regions were the economically weaker regions located in the periphery of the economically developed core regions (Copus, 2001). Nowadays, in the midst of the global economic crisis, there are rapid changes in the development perspectives of peripheral and core regions resulting to a more complex spatial relation between them.

The improvement of accessibility is the main instrument for overcoming peripherality and establishes the background for balanced regional development by enhancing the territorial cohesion of regions (Vickerman, 2004; Spiekermann & Wegener, 2006b). However, the effect from the improvement of accessibility on regional development depends on the region's initial level of competitiveness, as the region should be able to build upon the potential of this improvement (OECD, 2002; Venables & Gasiorek, 1999). Furthermore, it is considered essential to adopt the appropriate inter-disciplinary policy framework which would support the process of transport infrastructure development in order to achieve a positive result from the accessibility increase (Nogues & Salas-Olmedo, 2009). In Table 2 there is an example of how different policy priorities may lead to different goals for the development of transport infrastructure and produce various effects on peripherality.

Table 2. Example of the effect of different development policy goals on peripherality through transport development

Goals of policy	Goals for transport development	Effect on peripherality
a) Socio-economic convergence	Increased mobility	• Exchange of resources, knowledge and experience
b) Balanced competitiveness	Improved connections	• Regional competitiveness • Access to new activities
c) Environmental sustainability	Decrease of external cost	• Safeguard of the natural environment and resources • Resilience of the built environment
d) Economies of scale	Decrease of transport and production cost	• Exposure to unbalanced competition • Pump or tunnel effects

Source: Own elaboration

More specifically, the goal of socio-economic convergence promotes mobility between regions, which favours the communication of people and ideas, whereas the main goal of balanced competitiveness refers to the development of the appropriate connections towards the international market in order to access new destinations and enhance regional competitiveness. Furthermore, according to the goal for environmental sustainability, the transport infrastructure should be well adjusted to the environment without generating irreversible pressures on the natural resources. On the other hand, the economies of scale aim for the development of the transportation system in a way that it would minimise the total cost and, in this process, the peripheral region may be exposed to unbalanced competition while regional development may be compromised by pump and tunnel effects.

In Figure 4 there is a presentation of the process and possible outcomes from the improvement of transport accessibility between a peripheral and a core region. Prior to the improvement of transport accessibility, there is an interaction between the peripheral and the core region. This interaction may be based on common financial and commercial interests, political and administrative relations and social, cultural and historical relations and it is the driving force for the improvement of accessibility between them.

Moreover, the regions have specific priorities for socio-economic development, either common or different according to their individual needs. The goals of sustainable development, competitiveness, cohesion and improvement of the quality of life are common for all regions. However, there may be differentiations between a peripheral and a core region as to the degree of achievement of these goals. Furthermore, there are priorities for the peripheral region which have already been achieved by the core region. In peripheral regions, it is often observed a gap in the mobility conditions, the integration within the region and the access to the international market. The core region aims to further increase its economic growth and inter-regional networking and expand its activity towards new destinations.

The priorities for the development of the regional transportation system are formulated accordingly. The transport infrastructure of the peripheral region is relatively degraded in terms of effectiveness and provides limited intra-regional and inter-regional connections, while the core region's transportation system is more effective and able to expand in a competitive and innovative way ensuring its sustainability. In addition, the aforementioned priorities are affected by the external conditions (e.g. relation with other regions, trends of the global market, technological innovations etc).

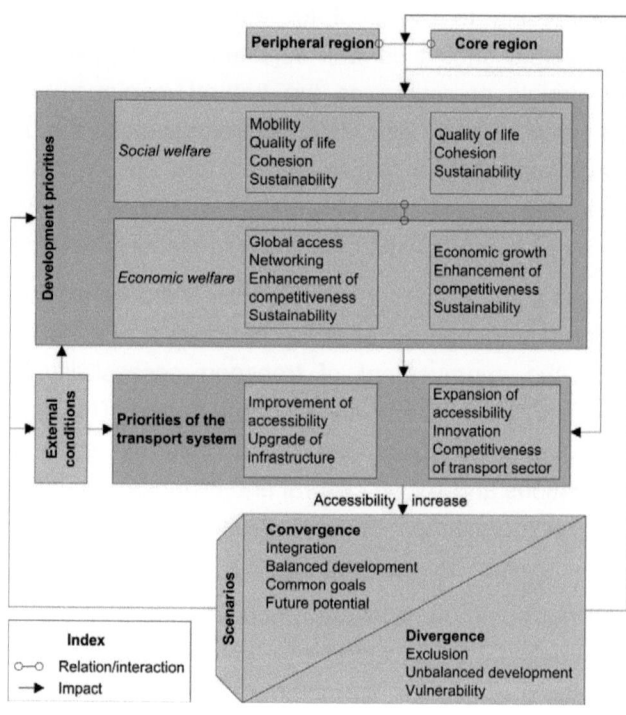

Source: Own elaboration

Figure 4. Potential impact of accessibility increase on a peripheral region and a core region

Under these conditions, there are many scenarios which may be realised from the increase of transport accessibility between a peripheral and a core region. These scenarios vary from the achievement of convergence, in terms of socio-economic integration and balanced regional development, which would allow the pursuit of common goals and future potentials, to the increase of divergence due to the unbalanced competition in favour of the core region, which will increase the vulnerability and exclusion of the peripheral region. The prevailing scenario will generate changes in the regions' relation and their development priorities affecting also the external conditions.

Consequently, the evaluation of the impact of transport accessibility on regional development and, more specifically, on the overcoming of peripherality depends on various parameters. According to Figure 5, in the case of the development or upgrade of infrastructure for the accessibility increase between two regions, the potential impact on peripherality should be evaluated through the global, the regional and the transportation perspectives.

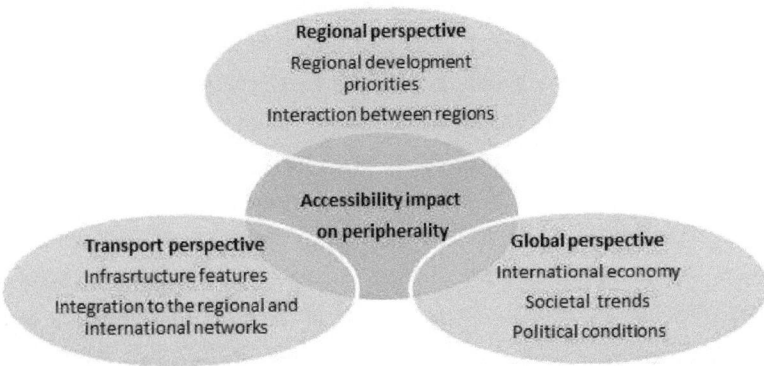

Source: Own elaboration

Figure 5. Parameters that influence the impact of the accessibility increase on peripherality due to the development or upgrade of transport infrastructure

The transportation perspective involves the features of the specific infrastructure, which should be evaluated in terms of the ability to improve the accessibility conditions. These features refer to the operational efficiency, the feasibility of the investment, the generated external costs and the innovative and competitive character of the project. Another main parameter is the integration of the infrastructure to the existing regional, inter-regional and international transport networks and the current or future operation of other, complementary or competitive, infrastructures.

The regional perspective focuses on the socio-economic conditions and the relevant priorities of the two regions in order to determine the required level of accessibility (Vickerman, 2003). The socio-economic conditions refer to the demographic characteristics, the economic activity, the productivity, the labour market, the availability of natural resources etc. Furthermore, the existing interaction between the two regions as well as the interaction with other regions should be taken into consideration.

From the global perspective, there are specific conditions in the international environment that may influence the impact of the improved accessibility on a peripheral region. These conditions depend on the needs of the international economy, the global societal trends and the general political conditions. During the Cold War period, for example, the potential for improving accessibility between the Western and the Eastern European regions would be considered irrelevant. Nowadays, this potential is a major priority of the European Union, but new obstacles emerge for its implementation due to the impacts of the global economic crisis. Another example concerns the progress of telecommunications and web based social networks, which contribute to the enhancement of networking among different regions and generate new needs for an improved level of transport accessibility between them.

3. EU POLICY FRAMEWORK FOR THE TRANSPORTATION SYSTEM AND ACCESSIBILITY

After the fall of the "Iron Curtain" and the end of civil warfare and political instability, the wider Balkan region entered a new era with the gradual accession of the Balkan countries to the European Union. The European policy framework considers the Balkans as a peripheral region and sets the process of its economic convergence and territorial cohesion as a main priority (European Commission, 2008a). The process of convergence faces now an additional challenge due to the global economic crisis which undermines the region's competitiveness. Against this pressure, the European Economic Recovery Plan aims to prevent a wide scale meltdown of the European economy due to the negative impacts of the crisis on the weaker member-states (European Commission, 2008b). At the meantime, a framework of common policy goals for the sustainable recovery of the European Union is structured based on the pillars of smart, sustainable and inclusive growth (European Commission, 2010).

Before the launch of the European enlargement towards the Balkans, the national economies of the region developed at a lower rate and, in most cases, in a different pattern in relation to the core region of Western Europe (Directorate General for Economic and Financial Affairs, 2005). The global economic crisis invaded these economies, while they were opening up towards the European and the international market. The process of European accession provided specific funding opportunities and policy reforms to these countries in order to upgrade their infrastructure and update their institutional frameworks. Thus, the countries managed to develop a series of socio-economic features which were essential for their defence against the crisis. However, there is still an active discussion as to if the national economies of Southeast Europe under institutional and socio-economic transition towards the European Union are more or less vulnerable to the pressures of the global crisis in comparison to the economies which maintain a more inward character (Bartlett & Prica, 2012).

In this changing environment, the development of the transportation system in the Balkan region and the enhancement of the accessibility conditions play a key-role in territorial cohesion and socio-economic convergence, which comprise main contributors towards the overcoming of the region's peripherality and the competitive development of the enlarged European Union (European Commission, 2008c). The purpose of this Chapter is to overview the main European policies for enlargement, territorial cohesion and transport development that formulate the priorities for improving the accessibility of the examined region.

3.1 Enlargement and convergence

In the period 1987-1996, the following countries of Central East and Southeast Europe applied successively for their accession to the European Union: Cyprus, Estonia, Hungary, Poland, the Czech Republic, Slovenia, Bulgaria, Latvia, Lithuania, Malta, Romania, Slovakia and Turkey. After the association agreements of 1997 with Turkey, Cyprus and Malta and the agreements with the countries of Central and Eastern Europe, the gradual accession of the aforementioned countries to the European Union has been decided according to the principle of differentiation and the degree of readiness of each country (http://europa.eu/legislation_summaries /enlargement/2004_and_2007_enlargement/e50017_en.htm accessed 13/11/2013).

In 1998 the European Council initiated negotiations with the candidate countries: Cyprus, Estonia, Hungary, Poland, the Czech Republic and Slovenia due to their high degree of preparedness. In 2000 the negotiations included the rest of the candidate countries excluding Turkey. In 2002 the Council agreed upon the readiness of these countries to fulfill the conditions for accession to the European Union. These conditions, also known as the Copenhagen criteria, refer to political stability, social equity, economic competitiveness and intention to adopt a common European legislation (European Council, 1993).

In 2003 the 5th European enlargement took place with the admission to the European Union of Cyprus, Estonia, Hungary, Poland, the Czech Republic, Slovenia, Latvia, Lithuania, Malta and Slovakia. The enlargement was concluded in 2004 with the admission of Bulgaria and Romania. The candidate countries became EU member states in 2007. (http://europa.eu/legislation_summaries/enlargement/2004_and_2007_enlarg ement/e50017_en.htm accessed 13/11/2013).

The negotiations with Turkey were initiated in 2005 while the involvement of Croatia with International Criminal Tribunal for the former Yugoslavia (ICTY) postponed the negotiations which were programmed for the same year. The accession of Croatia to the European Union was concluded in July 2013 (European Commission, 2013a). In the decade 2003 – 2013, thirteen new member states from the region of Central East and Southeast Europe accessed the enlarged European Union of 28 member states. Furthermore, the accession of Montenegro and Serbia is under negotiation while there is a group of candidate countries: the Former Yugoslav Republic of Macedonia, Turkey, Albania, Bosnia and Herzegovina and also Iceland. The accession of these countries would mean the accession of the wider Balkan region to the European Union.

In 2007 the first framework for the enlargement strategy was published by the European Commission (2007b) outlining the challenges and opportunities of the enlargement process. The framework is updated at a regular basis. The corresponding document for the period 2013-2014 addresses the issues of the progress of Croatia towards convergence to the European Union and the course of negotiations concerning the rest of the accession countries. It also highlights the need of economic and institutional reforms in the new member states and the accession countries as well as the necessity for the European Union to support the infrastructural upgrade of the Western Balkans against the pressures of the global economic crisis. In relevance to the transportation system, emphasis is given to the underdevelopment of cross-border transportation, the significance of infrastructure investments for the upgrade of accessibility between the Balkan region and the rest of Europe and the need for structural and competitiveness reform and standardisation initiatives for the region's transportation system (European Commission, 2013b).

3.2 Territorial cohesion

In the late 90s, the European Spatial Development Perspective, which aims to establish a common strategy in the European enlargement process, emphasises on the peripheral position of Southeast Europe and discusses the similarities and disparities within the region and between the region and the rest of the European Union in terms of the population, the economy, the environment and the transportation system (European Commission, 1999). In the context of transport development, the document sets as a main target the upgrade of infrastructure as well as the update of management and administrative methods in order to meet with the growing demands of the economy. At the meantime, the lack of financial resources and the limited regional perspective of the national policy are highlighted. The document outlines as priorities the enhancement of the transnational transport connections within the enlargement region and towards the rest of Europe, the convergence to a common transport policy and the structural reforms in rural areas.

Almost a decade after the European Spatial Development Perspective, the European Commission's Green Paper on Territorial Cohesion (2008c) was put into consultation in a very important period for the European Union's enlargement process, i.e. five years after the accession of twelve new member states from the wider Southeast European region and at the beginning of the global financial crisis. The document refers to the high rate of growth in the new member states over the past years. It focuses on the necessity to achieve a balanced development between the European regions through coordinated actions in order to take advantage of the territorial diversity and variety within the European territory. There are three dimensions in this course of action: a) The spatial dissemination of development in a balanced and sustainable way against the concentration of development in congested areas, b) the improvement of connectivity throughout the European regions by improving intermodal accessibility as well as high quality services and communication networks and c) the enhancement of cooperative inter-disciplinary networks from the regional to the European and international level. The document also highlights that the transport policy has a direct impact on territorial cohesion due to its contribution to the organisation of economic activity and settlements.

In 2011, the European ministers of spatial planning and territorial development agreed upon the Territorial Agenda of the European Union 2020, which presents the spatial perception of the Europe 2020 strategy (Informal Ministerial Meeting of Ministers responsible for Spatial Planning and Territorial Development, 2011). The document emphasises once again on the territorial diversity of the European Union and the prospect of achieving balanced development by exploiting the opportunities that this diversity offers while building effective cooperative networks between the regions. It also highlights the need to conduct common efforts in order to support the peripheral regions and develop balanced inter-regional relations. As main challenges in the development of the territorial cohesion are considered the global financial crisis and the reforms that should be conducted in the next period towards the social inclusion, the economic convergence and the preservation of the natural and cultural heritage and against the global climate change and the increased energy dependency of the European Union. The main goal in order to overcome these challenges is the polycentric regional development and the spatial integration of European institutional, policy and strategic planning frameworks.

3.3 Transport

The main European transport network which connects the Balkans to the rest of Europe is the trans-European transport network (TEN-T) with the purpose of promoting the balanced development and the intermodality of the transportation system, providing high quality and cost-efficient transport services, exploiting and upgrading the existing infrastructure and achieving an efficient level of accessibility throughout the European territory (European Commission, 2007c & 2009). In the Balkan region, the trans-European transport network provides links to the pan-European corridors which comprise the main European routes towards the East while aiming to promote the territorial cohesion of Southeast Europe (Balkan Task Force, 2000). The TEN-T priority projects are updated at a regular basis in order to meet with the needs of the EU transport policy. The most recent update was conducted in 2013 with the insertion in the priority projects of main European ports with rail and road links, airports with rail connections into major cities, railways to be upgraded into high speed rail and cross-border projects in order to improve East to West and North to South connectivity throughout Europe (ec.europa.eu/transport/themes/infrastructure/news/ten-t-corridors_en.htm accessed 27/11/2013).

Another recent initiative refers to the Comprehensive Network of the Western Balkans. The Comprehensive Network is managed by the Southeast Europe

Transport Observatory, which is an organisation established in 2004 in the context of the Memorandum of Understanding among the European Commission and the representatives of national transport authorities in the candidate member states: Albania, Bosnia and Herzegovina, Croatia, the Former Yugoslav Republic of Macedonia, Montenegro and Serbia. The scope of the Comprehensive Network is to strengthen the connectivity of the Western Balkans towards the TEN-T and TINA axes (SEETO participants, 2012).

The first attempt of the European Union to compose a common transport policy is the 1992 White Paper followed by the common transport policy Action Programme 1995 -2000 (European Commission, 1992 & 1995). The document mainly focuses on the establishment of the framework for the free competition in the transport market. In the next period, the uneven spatial distribution of development in the European Union, which compromised the goal for a sustainable transportation system, has led to the development of the 2001 White Paper (European Commission, 2001b). The document investigates the deficiencies of the transportation system in Europe, i.e. the unbalanced development of transport modes, the congested segments and nodes in the multimodal transport network and the negative impacts on the safety of people and goods, the public health, the social inclusion, the economic competitiveness and the preservation of environmental resources. In this context, it sets a series of sustainability objectives referring to the promotion of the combined use of transport modes, the decoupling of transport from economic growth, the investigation of alternative energy resources, the promotion of the free competition in the transport market and the investigation of investment possibilities for transport infrastructure of major significance for the European territorial cohesion. The White Paper emphasises on the fact that the institutional reform and the upgrade of intermodality are tools for the overcoming of the peripherality of rural regions. Concerning the wider Balkan region, the document focuses on the completion of the motorway network in Southeast Europe and its interconnection to the European networks, the development of the Motorways of the Sea (MoS) as part of intermodal routes through the Mediterranean Sea and the upgrade of the main international terminals as well as the enhancement of their hinterland connectivity. The goal for the development of a sustainable transportation system is also discussed in the 2011 White Paper, which highlights that many of the previous obstacles remain while new challenges are generated by the conditions of the global economy and the enlargement of the European Union (European Commission, 2011b). The document quantifies the new goals for transport sustainability for the next period while it concentrates on the strategic selection of transport investments with the criterion that they will contribute to the overall convergence and competitiveness of the European economy and on the energy efficiency of the transportation system.

The above review refers to the common transport policy framework of the European Union. Moreover, there are specific initiatives undertaken by the European Commission and other international organisations in cooperation with the candidate countries of the wider Balkan region. Before the 5th European enlargement, there has been an international focus on the monitoring and evaluation of the needs of the Balkans' transportation system and the development of the appropriate strategies for the transport sector. In the context of the 1999 Stability Pact, the Infrastructure Steering Group (ISG), represented by experts from the European Commission, the World Bank, the European Bank for Reconstruction and Development, the European Investment Bank and the European Development Bank, outlined a series of strategies for the improvement of the communications, energy and transport infrastructure in Southeast Europe (Busek & Kuhne, 2010). A more recent initiative refers to the European Commission's document entitled: "The EU and its neighbouring regions: A renewed approach to transport cooperation", where specific priorities for the multimodal transportation system of the Western Balkans and other EU neighbouring regions are set (European Commission, 2007d). Regarding the road transportation system, the document deals with the obstacle of delays and procedures due to border crossings, while it highlights the importance of converging to the international standards for vehicle safety and environmental performance. There are no exclusive measures regarding rail transportation in the Western Balkans as the main rail routes for freight transport between the European Union and its neighbours are located at Central and Eastern Europe (through Belarus, Moldova and Ukraine). Regarding the inland waterways, the main priority is the upgrade of infrastructure and management of inland navigation in the

Balkan region (Bosnia and Herzegovina, Croatia and Serbia) in order to support the full exploitation of the Danube region. In relevance to air transport, it is highlighted the importance for the strengthening of the comprehensive aviation agreements in order to expand the Single European Sky initiative. In the maritime system, it is proposed the convergence to the European Union's standards and policies and the promotion of the market liberisation with focus on the formulation of a free trade area in the Mediterranean. Emphasis is given by the European Commission on the prioritisation of transport projects and the allocation of investments (European Commission, 2007e). Finally, it should be mentioned that, in 2008, the European Commission proposed a Transport Community Treaty with the governments of the Western Balkan countries. The Treaty, which is under negotiation, focuses on the gradual integration of the region's transportation system to the European Commission in the fields of competitiveness, safety and sustainability http://ec.europa.eu/transport/themes/international/regional cooperation/caucasus_central-asia_en.htm accessed 15/11/2013).

3.4   Synthetic analysis

The synthetic analysis of the EU sectoral policies outlines the common priorities for the development of the transportation system in the wider Balkan region:

a) Establishment of transnational cooperative agreements and projects.
b) Investigation of funding prospects and efficient allocation of infrastructure investments.
c) Implementation of the necessary reforms for market liberisation regarding all transport modes and mainly air, rail and maritime transportation.
d) Facilitation of cross-border procedures and infrastructure.
e) Adoption of the EU standardisation framework in network development and system management.
f) Balanced development of transport modes and promotion of intermodality.
g) Specific priorities of high added value for each transport mode, i.e. completion of essential motorway connections, promotion of the Motorways of the Sea in the Mediterranean and development of the inland waterways in the Danube region.

These priorities will contribute to the development of the Balkan region's transportation system on the basis of three main policy pillars with different spatial reference:

- Regional. The regional pillar refers to the overcoming of peripherality through the polycentric development of the region and the exploitation of the local diversities.
- European. The main objective is to safeguard the smooth accession of the candidate countries in the Western Balkans and the gradual integration of the new member states to the European Union.
- International. The overall aim is to enhance the Balkan region's position in the international market while strengthening the enlarged European Union's international competitiveness.

There can be identified a direct or a potential impact of each policy priority for the Balkan region's transportation system on the three main pillars of the European Union's policy. The type of impact for each priority is described in Table 3, where the solid colouring corresponds to an impact targeted directly towards the specific pillar while the gradient colouring corresponds to an impact which can potentially support the specific pillar.

Table 3. Direct and potential impacts of transport policy priorities for the Balkan region on the main EU policy pillars

| | Impact on | | |
Priority	Polycentric regional development	EU enlargement and integration	International competitiveness
Transnational cooperation			
Infrastructure investment			
Intermodality			
Market liberisation			
Cross-border transport facilitation			
Motorways of the sea			
EU standardisation			
Motorway connections			
Danube inland navigation			

Source: Own elaboration

More specifically, the cooperation between the countries of the Balkan region with the EU core member states and the neighbouring countries in order to find mutually beneficial ways to develop and manage their transportation system will create development corridors and gateways that are able enhance the polycentric development, the European territorial and socio-economic cohesion and the international competitiveness. The same effect is expected from the strategic allocation of investments for the development or upgrade of transport infrastructure.

As it is described in Chapter 2, the combined use of transport modes for door-to-door mobility of persons and goods comprises a worldwide common practice in order to cope with the segregation of rural areas and to develop global connections in the context of the international market. In this way, the promotion of intermodality in the Balkans can generate a positive impact on the polycentric development, the EU integration and the international competitiveness.

The liberisation of the transport market in the Balkan region according to the European Union's specifications will promote the institutional convergence to the EU core region. Furthermore, it will formulate the conditions in order for the region to reach new destinations and expand its international potential. The polycentric regional development can be promoted through this process only in the case that it can contribute to the liberisation process.

The inward perspective of national policies before the accession, the remnants of the recent socio-political instability and the co-existence of the EU member states with other countries within the Balkan region are factors that increase the complexity of cross-border procedures and the lack of upgraded transnational corridors. The focus on the facilitation of cross-border transportation will improve the accessibility of the region towards Europe and international destinations. In this process, there is the opportunity for the emergence of new cross-border destinations, e.g. international transport terminals, work places, settlements, commercial areas, touristic destinations etc., which will promote the polycentric regional development.

The development of the Motorways of the Sea in the wider Mediterranean area will create intermodal connections from east to west through the Balkan region strengthening the territorial cohesion and the international accessibility of the enlarged European Union. At the regional level, the implementation of the appropriate local and regional strategies could turn the international seaports of the MoS into gateways that would support the socio-economic development of their hinterland leading to polycentric development.

Over the last two decades, the European Union proposes common standards and specifications for the transportation system in terms of strategic planning, development, organisation and management. The convergence to these standards and specifications will enhance the overall convergence of the Balkan region to the European Union. However, in order for them to contribute towards the region's polycentric development and international competitiveness, their adjustment and integration to the regional needs and diversities is required.

The completion of the Balkan motorway network is essential for the accessibility towards the rural areas of the region as well as for the interconnection with the motorway network of Western Europe and the corridors towards the East. On the other hand, the restoration of the inland navigation system in the Northern Balkans and along the Danube region will provide new development corridors within the region while integrating the main European inland waterways. These priorities can promote the region's international competitiveness with the precondition that these networks will become part of a comprehensive intermodal network according to the previously mentioned EU policy priority.

Summarising the analysis, it should be highlighted that the pillars of polycentric regional development, European integration and international competitiveness comprise the foundation of the contemporary EU development strategies. These pillars are the product of a gradual process which is based on the long term experience and the common efforts of the EU core member states. Consequently, they comprise essential components for the continuation of their common course towards competitive and sustainable development. On the other hand, the on-going European enlargement includes mainly the wider Balkan region, which presents much diversity within its territory and with the European core region in terms of historical and cultural background, socio-economic features and territorial characteristics. Moreover, the convergence of the peripheral region's new member states and the accession of the candidate countries coincide with the global financial crisis, which continues to put pressure on the national economies. Thus, the following questions can be asked:

- Can the same EU policy pillars be implemented for the development of a core and a peripheral European region?
- How can these pillars be adjusted to support the current and future development of the peripheral Balkan region?
- Which are the specific strengths that the Balkan region should depend upon in order to achieve convergence to these pillars while taking advantage of the specific competitive advantages of the region?

In order to answer these questions, a clear view of the differentiations and compatibilities between the Balkans and the EU core region should be acquired. Towards this purpose, the following Chapter involves a SWOT analysis based on the comparison between the transportation system of the Balkan region and the European Union.

4. ANALYSIS OF THE BALKAN REGION'S TRANSPORT SYSTEM

The scope of this Chapter is to identify the differentiations but also the similarities among the peripheral Balkan region and the rest of Europe. Moreover, the potential of the peripheral region is evaluated by highlighting the compatibilities and incompatibilities with the European policy priorities as well as the conditions of international competitiveness. The overall aim is to analyse the Balkan region's transportation system in the context of the on-going process of the European Union's enlargement, socio-economic convergence, territorial cohesion and transport development, which are the main sectoral EU policies that affect the development strategies of regional accessibility.

In order to analyse the transportation system of the Balkan region and investigate the features of peripherality and competitiveness in relation to the EU core region, a SWOT (Strengths – Weaknesses – Opportunities – Threats) analysis is conducted. The SWOT analysis was developed and conducted for the first time in the period 1960-1970 by the Stanford Research Institute for the Fortune 500 companies as a structured planning method for business management and market strategies (Picton and Wright, 1998; http://forlearn.jrc.ec.europa.eu/guide/4_methodology/meth_swot-analysis.htm accessed 30/11/2013). The principles and the methodology of the SWOT analysis have been adjusted and implemented in various scientific fields including the strategic planning purposes of the regional science (Karppi, et al., 2001)

4.1 Methodology of the SWOT analysis

The composition of the SWOT analysis for the Balkan transportation system is presented in Table 4.

Table 4. Composition of the SWOT analysis for the Balkan transportation system

Strengths	Opportunities
▲	▲
Competitive and lagging features of the Balkan region in relation to the EU core	Conditions and trends concerning the European and global environment
▼	▼
Weaknesses	Threats

Source: Own elaboration

More specifically:

- Strengths refer to the advantages of the regional transportation system regarding the provision of infrastructure and the generated transport task. They also include other regional features which generate a direct added value to the regional transportation system.
- Weaknesses are the disadvantages in the supply and demand features of the region's transportation system as well as other regional features with a direct negative impact on the transportation system.
- Opportunities comprise the overall conditions and trends which favour the competitive development of the transportation system and the regional accessibility conditions.
- Threats concern the regional and international factors which are likely to generate pressures on the current development course of the region's transportation system.

The first part of the analysis aims at the description of the results and findings based on the relation between transport and regional development and the review of the relevant EU policies, as presented in Chapters 2 and 3 respectively. The description is supported by the quantitative and qualitative analysis of the appropriate indicators, which derive from:

- The European Union's official statistical databases, i.e. Eurostat Transport Database
- The Enlargement Countries Database
- The publication: Panorama of Transport (Eurostat 2009) and
- The publication: Energy, transport and environment indicators (Eurostat, 2012).

The indicators have been processed for the purpose of the current manuscript and other recent projects of the author, i.e. the author's PhD thesis (Gavanas, 2011) and the papers: "Description of the new member states transport system in an era of convergence – Development of an indicator system" (Gavanas & Pitsiava, 2011) and "Methodology for the development of an integrated transport accessibility model for the wider Balkan region" (Gavanas & Pitsiava-Latinopoulou, 2013).

The SWOT analysis is concluded with the synthetic evaluation of the findings in order to present the comprehensive framework of the Balkan region's transportation system.

4.2 Findings of the SWOT analysis

4.2.1 Strengths

The Balkan region extends along the geographic border between Western Europe, Eastern Europe, Asia and Northeast Africa. Through history, its strategic location has generated opportunities for regional development but also external pressures by major international players (Grigor'ev & Severin, 2007). In the era of the European enlargement, the region's location is turning into a competitive advantage for the whole of the European Union, with the aim to be fully capitalised by the development of the Balkans as a hub for international trade (DTZ Research, 2008).

The region's transportation system is a main contributor towards this direction. The spatial distribution of the European transport network throughout the region provides a competitive advantage in terms of accessibility to the international market. The Pan-European axes of the region combined with the TEN-T network and complemented by the SEETO network connect the continents of Europe and Asia by land. The seaports along the Balkan coastline are connected to the European Union's Motorways of the Sea in the Mediterranean and Adriatic Seas and provide access to the Black Sea region (Gavanas & Pitsiava, 2011). The geographic location of the main European rivers provides a physical link from East to West and from North to South throughout the European territory (Eurostat, 2009). Moreover, the region's location strengthens its role in the energy distribution network between Europe and Asia, especially with the increasing energy dependency of the European Union (Eurostat, 2012).

The relatively high density of the railway network in the Northern Balkans, deriving mainly from the development priorities before the EU accession era, is a significant feature of the regional transportation system. More specifically, the railway density over surface in Slovenia, Hungary and Romania is over 10 km / 100 km^2 whereas the corresponding average density in the enlarged European Union (EU27) is less than 8 km / 100 km^2. The corresponding values for railway density over population are over 11 km / 10,000 inhabitants while the EU27 average is approximately 8 km / 10,000 inhabitants (Gavanas, 2011).

In the past decades, the Balkan railways were usually used for freight transportation. In 2008, the average railway freight task for the twelve new member states accessed in the 5th European enlargement (EU12) was 15,037·10^6 ton-km (with a slight decrease over the period 2004-2008) in comparison to an average of 17,836·10^6 ton-km for the fifteen pre-enlargement member states (EU15) (Gavanas, 2011). The major freight transporters by rail among the new member states are Poland, Latvia and the Czech Republic, which are not considered as part of the Balkan region by the specific research. Within the Balkan region, the countries with the highest task are located at the North of the region, i.e. Romania (15,236·10^6 ton-km) and Hungary (9,874·10^6 ton-km) (Gavanas, 2011).

4.2.2 Weaknesses

In spite of the progress in the restoration of the motorway network in the Balkan region, which is a main priority of the trans-European network, the pan-European axes and the SEETO network, it is observed an unbalanced spatial distribution of motorways within the region. More specifically, the high density of motorways in the EU core region seems to affect the availability of motorway infrastructure in its adjacent Balkan countries. According to data for 2010, the density over surface for the Slovenian motorway network was 3 km / 100 km^2, which was equal to two times the EU27 average and four times the density of the Romanian motorway network (0.07 km/100 km^2) (Gavanas & Pitsiava-Latinopoulou, 2013). An even bigger gap is observed when one examines the availability of motorway infrastructure in relation to the population of each country. In 2010, the motorway density over population was 3.7 km / 10,000 inhabitants for Slovenia while it was only 0.16 km / 10,000 inhabitants for Romania (http://epp.eurostat.ec.europa.eu/portal/page/portal/ transport/data/database accessed 23/11/2013). The differentiation in the provision of motorway infrastructure leads to a similar differentiation in transport demand. In 2008, the road traffic task along the national motorway networks of Slovenia was $4.88 \cdot 10^6$ vehicle–km, over 3.5 times higher than the task of Romania ($1.36 \cdot 10^6$ vehicle–km) (http://epp.eurostat.ec.europa.eu/portal/page/portal/transport/data/database accessed 23/11/2013).

A very low passenger task is observed in the Balkan railway system in comparison to the EU core region due to the lack of upgraded infrastructure and competitive services. Moreover, there is a significant differentiation within the Balkan region with the railway passenger task ranging from $9,617 \cdot 10^6$ passenger-km for Serbia, $9,524 \cdot 10^6$ passenger-km for Hungary and $8,092 \cdot 10^6$ passenger-km for Romania to $2,422 \cdot 10^6$ passenger-km for Bulgaria, $1,811 \cdot 10^6$ passenger-km for Greece and $724 \cdot 10^6$ passenger-km for Slovenia. The comparison to the highest passenger tasks of the EU core region, i.e. $79,835 \cdot 10^6$ passenger-km for France and $78,735 \cdot 10^6$ passenger-km for Denmark, is indicative of the limited role of railways in the total passenger transport service (Eurostat, 2009).

Regarding the inland navigation system, there has been a very small increase in the length of inland waterways in the European territory over the past decade. In 2010, the Balkan countries with the longest navigable waterway networks were Romania with 1,779 km and Hungary with 1,587 km, while the length of the German navigable waterways was 7,728 km (http://epp.eurostat.ec.europa.eu/portal/page/portal/transport/data/database accessed 23/11/2013). The freight transport demand of inland waterways in the Balkans is mostly concentrated on national trips (Gavanas and Pitsiava, 2011; Gavanas, 2011).

The lack of effective land connections diminishes the role of the region's seaport network in the European Motorways of the Sea in the Adriatic, the Mediterranean and the Black Seas. According to the data of 2010, the total volume of goods transported through the seaport network of the whole of the

Balkan region (excluding Montenegro due to the absence of reliable data) was approximately $228.5 \cdot 10^6$ tons, with the Greek seaports accounting for more than 50% ($124.4 \cdot 10^6$ tons). This volume is equal to the volume which was transported in the same year by the seaport network of Belgium ($228.2 \cdot 10^6$ tons), which is a country with access to the North Sea through a total of four main seaports (Gavanas & Pitsiava-Latinopoulou, 2013). In addition, the number and capacity of container handling services at the Balkan seaports is relatively small (Gavanas, 2011).

The international airport infrastructure and services in the region is still relatively underdeveloped in comparison to the EU core region. More specifically, the EU27 average number of main airports per country, i.e. airports that handle more than $15 \cdot 10^3$ passenger movements annually, is 10, while the corresponding number ranges from 1 to 5 in the Balkans, with the exception of Greece (19 main airports) due to the presence of a large number of remote and/or touristic destinations in the Greek islands (http://epp.eurostat.ec.europa.eu/portal/page/portal/transport/data/database accessed 2/12/2013).

4.2.3 Opportunities

The recovery of the Balkan region from the recent socio-political turmoil followed by a period of stability and cooperation within the region and with the neighbouring countries contributes towards the restoration of missing inter-regional transport links and the enhancement of the region's international accessibility. Moreover, they formulate the conditions to integrate these cooperative initiatives into a sustainable regional transportation system by building upon its advantages and understanding its drawbacks and specific requirements.

The European Union's enlargement process towards the Balkan region continues to ensure funding options and institutional support for the upgrade of transport infrastructure in the Balkans in order to promote the cohesion and competitiveness of the whole European territory. Moreover, the European Union's past experience of over a decade in the sustainable development of the transportation system and the strengthening of regional competitiveness through the enhancement of accessibility has produced a significant inventory of strategies, tools, methods and good practices which are available to the Balkan region's decision makers and stakeholders.

In this context, the geography and location of the Balkan region in combination with the strengths of its transportation system are favouring the development of a multimodal land transport network with strong links towards the railway and inland waterway networks of the North and the seaport network of the South. Furthermore, the multimodal land transport network should focus on the enhancement of passenger services by rail and the exploitation of the potential of the Danube region. Other priorities refer to the strategic upgrade of intermodal land networks as a measure against the uneven distribution of the motorway network within the region. The seaport network can enhance its global competitiveness through the modernisation of its freight handling services and the capitalisation of its locational advantage. In addition, the intense competitiveness of the air transport market and the increasing attractiveness of the Balkans as a destination are expected to lead to major improvements for the region's airport network.

4.2.4 Threats

The main threats for an efficient and competitive regional transportation system in the Balkans derive from the global economic crisis. The impacts of the crisis on transport development can be direct or indirect. The direct impact refers to the scarcity of funding resources for the upgrade of transport infrastructure and services. The decrease in funding may lead to major barriers in the effort of exploiting the current strengths and the opportunities presented for the region. On the other hand, the indirect effect of the crisis refers to the potential re-evaluation of the development priorities at the national and European level and the shift in the balance of the global market, which may affect the development strategies of the Balkan transportation system.

The different organisational and operational features of the transportation system and the peripheral role of the Balkan region in the European and international environment obstruct the formulation of a common transport policy framework for the enlarged European Union. Moreover, the lack of a common strategic approach by the decision makers of the Balkan region creates a gap between the overall priorities of the enlarged European Union and the goals set by the national policy framework. Despite the undisputable progress over the last decade, the indicator results presented in the S.W.O.T. analysis suggest that there is still a considerable distance to be travelled towards the effective convergence of the transportation system in the European territory.

4.3   Synthesis of the SWOT findings

The socio-political stability and the increase of economic development over the last decade in the Balkans are leading to the increase of the regional mobility needs and of the demand for reaching new destinations throughout the region. The competitive geographic location of the Balkans and the gradual accession to the European Union are assets for the development of an effective multimodal transportation system with the aim to service the increasing transport demand of the region.

The transportation system of the Balkans has inherited some competitive features from the period prior to the EU accession, such as the relatively high railway density and the structure of its seaport network, while significant international and regional infrastructure is being developing during the accession period, such as the expansion of the TEN-T network.

The examination of the features of transport demand concerning the Balkan region shows specific deficiencies in relation to the corresponding features of the European Union. This fact highlights the peripheral role of the region, which has not yet been overcome by the aforementioned development of the region's transport infrastructure. In addition, the indicators highlight the unbalanced development between different transport modes in terms of the network distribution and the traffic flows within the region, which is opposed to the EU policy priorities for sustainable transportation.

An additional challenge in the process of developing a sustainable and competitive transportation system in the Balkan region derives from the global economic crisis, which seems to gradually acquire a more permanent role in the setting of priorities at the national and the European level. Apart from the evident scarcity in available resources for the upgrade of the regional transportation system, the shift in the balance of the global market is expected to generate incompatibilities and conflicts regarding the priorities of regional development which may lead, in the long-term, to another period of instability in the region.

The enhancement of the common regional perspective regarding the strategic planning of the transportation system in a way that it would integrate the national policies into a competitive European and international framework would contribute to the confrontation of the aforementioned challenges. The development of the appropriate tools for the monitoring of the main transport features in a comprehensive way for the wider Balkan region is an essential step towards this direction. The aim of the following Chapter is the development of such a comprehensive framework for the formulation of a specialised model for the monitoring of transport accessibility in the region.

5. METHODOLOGY FOR THE DEVELOPMENT OF AN INTEGRATED ACCESSIBILITY MODEL FOR THE BALKAN REGION

As it is discussed in the theoretical approach of the manuscript, accessibility comprises the common ground of interaction between the transportation system and the factors of regional development. A high level of accessibility can potentially contribute towards regional development, depending on the readiness of the region to take advantage of this increase. A low level of accessibility is expected to worsen the peripheral position of the region in relation to the international market. Thus, it is essential to develop a tool for the monitoring of the accessibility conditions in relation to the features of regional development.

Based on the policy goals and the transport related features presented in the previous Chapters, the scope of the current Chapter is to define the components for the formulation of the appropriate model for the assessment of accessibility for the Balkan region and, in specific, the accessibility of the region's main poles of socio-economic development towards the transport gateways of international significance. The first section refers to the presentation of the structure of regional accessibility models followed by a brief overview of the main accessibility models implemented for the European regions. In the next section, the specific components for the Balkan region's accessibility model are proposed and the main parameters for the development of the appropriate implementation scenarios are presented.

5.1 State of the art in Europe

The investigation of the international literature in the subject of regional transport accessibility models shows a wide range of practices using different patterns of analysis, measurement units or even conceptual approaches. The differentiations between the available models are due to the following factors (Gavanas & Pavlidou, 2011; Knowles, et al., 2008):

- Accessibility and its significance for the formulation of the conditions for regional development are relatively recent concepts.
- The components of an accessibility model depend on the scope of the study, the geographical reference, the spatial level of the analysis as well as the purpose, type and mode of transport.

- The level of accessibility cannot be assessed through direct observation and measurement and its calculation requires an inter-disciplinary approach of transport, spatial and socio-economic features.
- The combination of features from different disciplines implies a dynamic relation between them. For example, the level of accessibility of an origin towards a set of destinations through the motorway network may be increased by the upgrade of a motorway connection or decreased by the national policy for reducing roadway emissions.
- The above factors create obstacles in the standardisation of the methodological procedures while the integration of the produced geospatial databases is difficult. A useful tool against these obstacles derives from the application of contemporary Geographic Information Systems (GIS).

Despite the aforementioned differences, the accessibility models are based on the following function (Schurmann, et al., 1997):

$$A_{i\{1,m\}} = \sum_{j=1}^{n} G\left(F_{ij}, L_j\right) \qquad \text{Equation 1}$$

Where

$i, j =$ Node of origin and destination respectively, i.e. urban centres, transport nodes, centroids, raster cells etc.

$F_{ij} =$ Function of friction or impedance (as an expression of time-distance, cost etc.) for reaching the node of destination (j) from the node of origin (i).

$L_j =$ Function of activity (as an expression of a socio-economic feature such as population, income, GDP etc.) conducted in the node of destination (j) and attracting trips from the node of origin (i).

There are three general categories of accessibility model expressions (Rodrigue, 2013; Spiekermann & Neubauer, 2002): a) Travel cost models for the assessment of the total or average time-distance or cost for travelling from the origin to the pre-selected set of destinations, b) Daily accessibility models for the estimation of the destinations that can be reached using a "fixed budget", e.g. a total of available time and c) Potential accessibility models, where the activity conducted in the destination is used to weight the attraction of trips from the origin. Due to the inclusion of the socio-economic activity as a main component of the model, the potential accessibility models are often used to support decisions in the fields of strategic planning and policy making. In this context, the current research focuses on the specific category.

A comprehensive and useful review of the main accessibility models implemented in the European territory is presented by the European Spatial Planning Observation Network (ESPON) in the report on "Transport services and networks: Territorial trends and basic supply of infrastructure for territorial cohesion" (ESPON, 2003). According to the report, Keeble et al. (1982; 1988) developed the first model in Europe for the assessment of the potential accessibility of NUTSII centroids of the 9 countries that comprised the European Communities at that time and the 12 countries that later on created the European Union. The set of destinations of Keeble's model is located inside and outside the European territory, while it is weighted by GDP. Since the early 90s, several models were developed covering various European regions, the EU12 or the EU15 area and also the wider European territory and referring either to passenger or freight movements by the multimodal or the road transport network, such as: (Bruinsma & Rietveld, 1993), (Gutierrez & Urbano, 1996), (Copus, 1999), (Diego, 2002), (Schurmann & Talaat, 2000), (Baradan, 2001). The aforementioned models differ also in their components and their mathematical expression.

One of the most commonly used regional accessibility models in the European Union is the model developed in the framework of the EU research programme SASI (Socio-economic and spatial impacts of the Trans-European transport network) in the period 1996-1999. The SASI model refers to a series of scenarios for the development of economic and demographic characteristics, infrastructure investments and the gradual completion of the trans-European network (Wegener, 2008). The model has a complex structure with a part that refers to the estimation of accessibility by taking into account the region's available transport links to the TEN-T, the regional socio-economic and demographic characteristics and the impact of the EU policies and strategies for the development of Europe's transportation system. The SASI model's function of friction is an expression of generalised cost for the travel between a pair of origin and destination within and throughout each European region by road, rail and air.

In the same concept, the ESPON's potential accessibility model is implemented for the monitoring of the impact of accessibility on territorial cohesion (ESPON, 2009). The mathematical expression of the model is exponential, the origins are NUTSIII centroids and the destinations are raster cells of 10 x 10 km. The attractiveness of destinations is weighted by population. The model is developed following similar patterns for the road, rail and air transport network. Furthermore, it is used for the assessment of multimodal accessibility. Using data from the implementation of the accessibility models for each mode, the multimodal accessibility model is based on a combined function of friction, according to which the sensitivity of the generalised cost of each mode is weighted by a parameter of sensitivity.

The review of European literature shows that there are no specialized models developed for the Balkan region. However, several of these models, including SASI and ESPON, cover the wider European territory extracting similar conclusions for the peripheral region of the Balkans. A representative example refers to the Guitierrez and Urbano's model for the estimation of the change in the accessibility level of the urban centres of the 12 EU member-states in 1992, after the completion of the first phase of the TEN-T network. The study focused on the peripheral EU countries (Ireland, Spain, Portugal and Greece) highlighting that the completion of the TEN-T's first phase was not expected to diminish the difference of accessibility between the "core" poles of Western Europe and the "peripheral" ones, as it would lead to a similar increase in the accessibility level of both (Guitierrez & Urbano, 1996). The ESPON maps regarding the accessibility levels in the European regions lead to a similar conclusion (ESPON, 2003).

5.2 Preliminary case study of accessibility change in the Balkan region

Primary to the presentation of the components for the formulation of a specialised potential accessibility model for the Balkan region, a case study is conducted in order to verify the changeability of accessibility within the region due to the development of the trans-European transport network and the progress of the EU enlargement process. In order to achieve this, a simplified accessibility model is developed and implemented. The model adopts the SASI and ESPON approach regarding the investigation of the impact of the overall EU policies and strategies on regional and local development[3]. The function of friction is based on the Connectivity Indicator of ESPON (2003) using the total travel time (including delays) as the expression of impedance between the origin and the destination. The mathematical expression is a modification of the Gutierrez and Urbano (1996) model.

5.2.1 Model description

The model refers to the road freight transport from the main urban centres to the region's international seaport network. In specific, the function of the simplified accessibility model is the following (Gavanas, 2011):

$$A_i = 100 \cdot \left(\frac{\sum_{j=1}^{n}\left(F_{ij} \cdot L_j\right)}{\sum_{j=1}^{n} L_j} \right)^{-1} \qquad \text{Equation 2}$$

Where

$i =$ City of origin.

$j =$ International seaport of destination.

$F_{ij} =$ Total travel time (including stops and delays) by truck for the access of the seaport of destination (j) from the city of origin (i).

$L_j =$ Annual handling of freight (in tons) by the seaport of destination (j).

The cities with population of more than $0.5 \cdot 10^6$ are included in the model, as the main economic and industrial centres of the region. In the case study, the cities of Belgrade and Sofia are selected as origin cities due to their position at the centre and south of the Balkans respectively.

---

[3] It should be highlighted that the case study was concluded in 2010 and, thus, does not include updated information, such as the 2010 TEN-T priority projects.

The international seaports of destination are selected under the condition that they are part of the region's seaport network but also linked to the EU Motorways of the Sea and, consequently, part of a multimodal route connecting the origin cities to the rest of Europe and the world. They should be classified as Category A seaports and handle freight of more than $2 \cdot 10^6$ tons / year. In the case study, the main seaports of the Adriatic and Ionian Seas are selected, i.e. Rijeka, Split and Ploce in Croatia, Bar in Montenegro, Durres in Albania, Igoumenitsa and Patra in Greece (Figure 6).

The function of friction, which measures the total travel time from origin to destination in hours, is described by the equation:

$$F_{ij} = \sum F_a + \sum F_n + \sum F_b (+ F_h) \qquad \text{Equation 3}$$

Where

$F_a=$ Impedance of road segments. It derives from the multiplication of the travel time (T_a) on a road segment of a specific class (according to the international classification) with a coefficient of safety and comfort (C_a) for driving on a road segment of the specific class. The average travel speed for the calculation of travel time and the coefficient of safety and comfort per road class are given in Table 5.

$F_n=$ Delay due to traveling through the urban or suburban network at the outskirts of main cities due to the influence of urban and suburban road traffic. It depends on the population of the city according to the equation: $I_n = 15 \cdot \log(P \cdot 10)$, where ($P$) is the population in 10^6 inhabitants (Guitierrez & Urbano, 1996).

$F_b=$ Delay due to border crossings, which is equal to an average of 2 hours only in the case that at least one of the two adjacent countries is not an EU member state. The penalty is calculated as an average of waiting times in border crossings given by the Research Programme: "European Space and Territorial Integration Alternatives Spatial Planning Observatory Network in Southeast Europe-ESTIA SPOSE" (INTERREG IIB/CADSES, 2004-2006).

$F_h=$ Obligatory break for truck drivers according to the driving schedule for professional drivers (ESPON, 2004).

Table 5. Average travel speed and coefficient of safety and comfort per road class

Class of road segment	Average travel speed (trucks)	C_α
Motorway	75 km/h	0.8
National road	55 km/h	1.0
Regional or other road	40 km/h	1.2

Source: ESPON, 2004; Gutierrez and Urbano, 1996

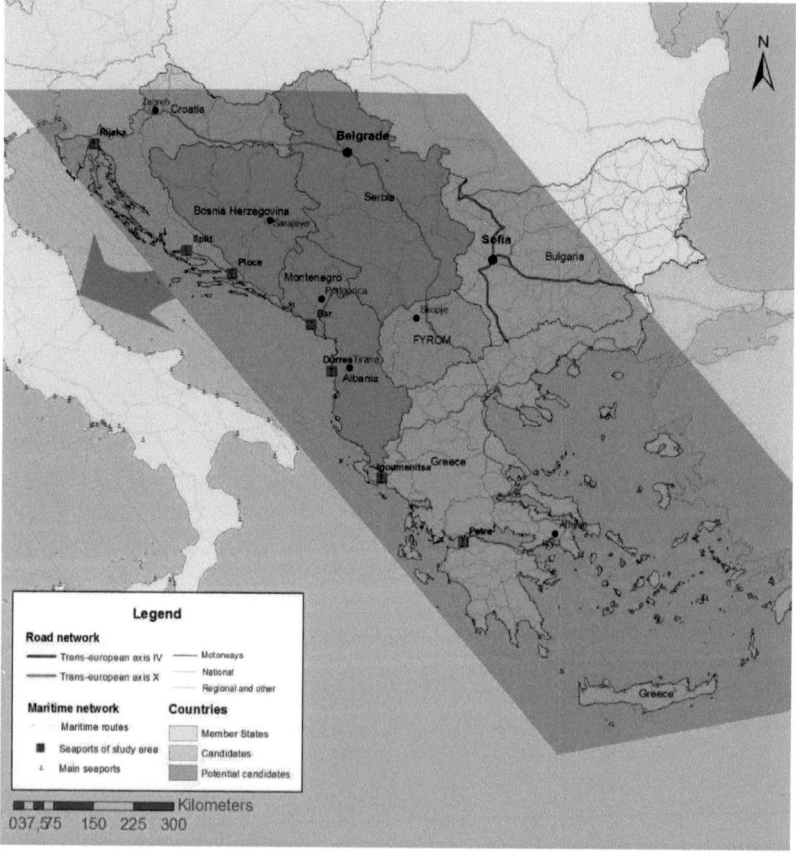

Source: Gavanas & Pavlidou, 2011

Figure 6. Study area

5.2.2 Presentation of results

A total of 5 scenarios are implemented for the evaluation of the change in time-distance and accessibility in the specific case study. The aim is to examine the tendencies in the region's accessibility due to the completion of the priority projects of the trans-European road network for the period 2005-2010 (European Commission, 2005) and the gradual accession of candidate and potential candidate member-states of the region in the European Union. In Table 6, the conditions which apply for each scenario are marked with the symbol (X). The time horizons of the mid-term and long-term scenarios are 2015 and 2020 respectively. In the long-term Scenarios 3 and 4, a factor of lagging is inserted either in the completion of the trans-European network or in the accession process as a result of the global economic crisis and its impact on the region and the wider European Union, based on the estimations of relevant literature (Centre for European Reform, 2008).

Table 6. Case study scenarios

Time reference	Scenario	Development of TEN-T			EU enlargement process		
		Before priority projects 2005	After priority projects 2005	Completion	27 member-states	Accession of candidates	Accession of candidates and potential candidates
Current	1	x			x		
Mid-term	2		x		x		
Long-term	3			x	x		
	4		x			x	
	5			x			x

Source: Gavanas & Pavlidou, 2011

In Table 7 there is a presentation of the estimated change rates in impedance (time friction) for the mid-term and long-term scenarios compared to the impedance value of Scenario 1. Furthermore, the accessibility change rate in relation to the value of accessibility for Scenario 1 is presented in Table 8. There are two results calculated for each connection of origin and destination in each scenario. In the first, the obligatory break for the resting of truck drivers is not taken into account, while in the second the corresponding time friction (F_h) is inserted in Equation 3.

Table 7. Impedance change rate per scenario in relation to impedance value for Scenario 1

Seaport	Impedance (h) Scenario 1		Change in impedance (%)							
			Scenario 2		Scenario 3		Scenario 4		Scenario 5	
	No break	With break	No break	With break	No break	With break	No break	With break	No break	With break
					Belgrade					
Rijeka	9.8	10.5	0.0	0.0	-14.1	-13.1	0.0	0.0	-34.6	-32.1
Split	17.1	26.8	0.0	0.0	-9.2	-3.1	0.0	2.8	-44.4	-61.0
Ploce	13.5	23.3	0.0	0.0	-10.3	-6.0	0.0	0.0	-54.7	-70.5
Bar	11.1	20.8	0.0	0.0	0.0	0.0	0.0	0.0	-18.1	-52.8
Durres	14.2	24.7	0.0	0.0	-6.3	-6.6	0.0	0.0	-34.3	-58.4
Igoumenitsa	16.5	27.0	-6.1	-3.7	-17.5	-10.7	-12.1	-7.4	-34.6	-23.9
Patra	19.9	39.4	-5.0	-25.4	-22.8	-34.4	-10.1	-27.9	-37.0	-43.4
					Sofia					
Rijeka	17.3	27.8	0.0	0.0	-14.6	-9.1	0.0	0.0	-37.7	-26.2
Split	22.5	42.0	0.0	0.0	-6.89	-3.7	0.0	0.0	-33.5	-39.4
Ploce	19.0	29.5	0.0	0.0	-13.0	-8.3	0.0	0.0	-44.6	-31.2
Bar	15.2	25.7	0.0	0.0	-7.6	-4.5	0.0	0.0	-33.8	-22.9
Durres	15.7	26.2	0.0	0.0	-37.8	-59.8	0.0	0.0	-63.2	-75.1
Igoumenitsa	9.2	9.9	-13.6	-12.6	-21.6	-20.0	-13.6	-12.6	-21.6	-20.0
Patra	12.6	22.3	-9.9	-5.6	-28.9	-56.6	-9.9	-5.6	-28.9	-56.6

Source: Gavanas & Pavlidou, 2011

Table 8. Accessibility change rate per scenario in relation to accessibility value for Scenario 1

	Accessibility (h⁻¹) Scenario 1	Change in accessibility (%)			
		Scenario 2	Scenario 3	Scenario 4	Scenario 5
		Belgrade			
No break	7.4	2.7	16.2	4.0	63.5
With break	4.7	5.1	17.0	8.5	91.5
		Sofia			
No break	6.1	1.6	19.7	1.6	62.3
With break	3.8	0.0	21.0	0.0	52.6

Source: Own elaboration

5.2.3 Conclusions from the case study

The main conclusions of the preliminary case study concern the following:

- In Scenario 1 (current situation), delays are presented due to border crossings between non EU member states.

- In Scenario 2, the TEN-T priority project which is included in the 2005-2010 agenda affecting the study area refers to the upgrade of the segment that connects Sofia with the Bulgarian-Greek border along the axis IV. The upgraded segment is expected to attract traffic from alternative routes but has limited effect on the overall accessibility conditions of the Adriatic/Ionian seaport network.

- On the other hand, the completion of the TEN-T road network, as it is examined in Scenario 3, will lead to a significant improvement of accessibility despite cross-border delays between non EU member states.

- The most interesting finding from the examination of Scenario 4 is the change in impedance which affects the shortest routes connecting Belgrade with the Greek seaports due to the accession of candidate member state F.Y.R.O.M.

- Finally, the results from the implementation of the long-term Scenario 5, with the completion of the TEN-T road network and the accession of the Balkan countries to the European Union, indicate a significant improvement in both the travel times and the overall accessibility conditions towards the seaport network of Adriatic/Ionian.

From the case study, it is evident that the development of major transport infrastructure and the simplification of the border-crossing procedures due to the accession of the countries in the Balkan region to the European Union are main contributors to the improvement of regional accessibility. It can be also concluded that there is a diverse impact on specific routes and on the relation of specific pairs of origins and destinations according to the progress of the TEN-T and the accession process.

The above analysis confirms that the potential accessibility model which was developed and implemented in the preliminary case study fulfils its goal to investigate the changeability of accessibility within the Balkan region as an output of the development of the TEN-T network and the enlargement process. However, it has specific drawbacks due to the lack of sufficient interdisciplinarity. More specifically, the routing is based on time distance and the conditions of safety and comfort without taking into account the elements of external cost. The external costs generated by the operation of the seaports of destination should be included in the measuring of their attractiveness. Moreover, the change rate of freight at seaports and of urban population is estimated as proportional to the change rate in the previous period. The accession to the EU and the global economic crisis are not taken into account in these estimations.

Thus, it is considered essential to construct a more specialised potential accessibility model through the interdisciplinary cooperation of scientists from different fields of expertise. The model should include comprehensively the necessary components within a flexible pattern that can be adapted to the local conditions and the features of the multimodal transportation system. Towards this purpose, the following Section provides a presentation of the appropriate components and methodology.

5.3 Components of an integrated potential accessibility model for the Balkan region

According to Equation 1, in order to develop and implement the integrated potential accessibility model for the Balkans, the following tasks are necessary:

- Selection of the set of origin nodes (i) and the set of destination nodes (j).
- Formulation of the function of friction (F_{ij}) and function of activity (L_j).
- Development of the implementation scenarios.

The successful completion of these tasks requires the cooperation of experts in the analysis of the availability of infrastructure and the trends in transport demand, the assessment of the features of socio-economic development and environmental sustainability related to the origin and destination nodes as well as the operation of the examined transport network, the statistical analysis of the model's data and outputs and the development of a geospatial database for calculations.

In the context of selecting the appropriate set of origins, it is proposed to select the main urban centres of the region, i.e. the administrative capitals and other major cities, as they concentrate the majority of the regional population and activity while comprising the landmarks of the region's international competitiveness. The international airport terminals and seaports are suggested as destination nodes for passenger and freight transport respectively, as they attract the region's population and goods towards international destinations. The railway and inland waterway terminals are not taken into consideration since their role in international transportation is limited in comparison to airports and seaports (see Chapter 4 of the manuscript). A description of the proposed nodes of origin and destination are presented in Table 9.

Table 9. Proposed nodes of Origin and Destination

Nodes of origin (i)	
Administrative capitals of regions and prefectures	
Urban and industrial centres selected by population and/or economic criteria	
Nodes of destination (j)	
Type of transport	
Passenger	*Freight*
Main international airports*	Main international seaports**
* According to Eurostat, airports that handle at least $150 \cdot 10^3$ passengers / year	
** According to Eurostat, seaports that handle at least $200 \cdot 10^3$ passengers / year or at least 10^6 tons / year	

Source: Own elaboration

In Table 10, there is a presentation of the components which are proposed for the formulation of the function of friction (impedance) according to the examined type of network, i.e. land, air/maritime and all. The estimation of cross-border delays is of special interest for all transport modes operating in the Balkans, as the region comprises both EU member-states and other countries with various cross-border conditions. Regarding the land transport network, it should be noted that many (mainly road) axes of regional and national importance in the Balkans are locally affected by peri-urban congestion due to the fact that they often extend throughout the outskirts of cities. Finally, the factors of internal cost (imposed on the immediate stakeholders) and external cost (imposed on the society and the environment) should be included in the calculation of the total friction in order to take into account both the purely financial aspect, which is crucial in the current recession period, and the aspect of sustainability, which is a constant priority for regional development. Furthermore, the overall travel cost (friction) in the case of a multimodal route linking an origin to a destination should be estimated as the sum of costs generated by traveling along each participating type of network, where each cost is weighted by the total task (in passenger-km or ton-km) produced along the specific network.

Table 10. Components of the function of friction (impedance)

Function of friction (F_{ij})		
Type of network		
Land	*Air / Maritime*	*All*
• Time-distance on segment • Delays due to peri-urban congestion	• Time duration of trip	• Delays at transport nodes (due to boarding, interchange, loading/unloading, transhipment etc) • Delays due to cross-border procedures
Weighted by the factors of: • Internal cost (operation and management) • External cost (energy consumption, emissions, land take, segregation, labour market, health etc)		

Source: Own elaboration

As it is suggested in Table 11, the function of activity that is conducted at the nodes of destination (main international airports and/or seaports) should be an expression of their international competitiveness and can be expressed by their annual international (passenger and/or freight) task. This should be weighted by the level of service and the external cost due to the terminal's operation, a process that provides to the model the ability to select the most attractive node of destination according to its size and quality of service as well as the impact of its activity on the factors of sustainability.

Table 11. Components of the function of activity

Function of Activity (L_j)	
Type of transport	
Passenger	*Freight*
International passenger-km per year	International ton-km per year
Weighted by the factors of: • Level of service (regularity, frequency, safety, comfort, coverage, intermodality, innovation etc) • External cost (energy consumption, emissions, land take, segregation, labour market, health etc)	

Source: Own elaboration

Finally, the implementation of the potential accessibility model requires the development of the appropriate scenarios. The geographical coverage as well as the spatial and time reference of the scenarios may differ depending on the scope and objectives of each study. Regardless the specific requirements of the study, there can be identified a set of components that should be taken into consideration as they affect the level of accessibility and the way that it can be exploited by the region in order to enhance the aspects of cohesion and competitiveness (Table 12). The components of Table 12 derive from the S.W.O.T. analysis of Chapter 4, where they are discussed in detail.

Table 12. Components of the implementation scenarios

Thematic field	Level of reference		
	Regional	*European*	*International*
Transport	• Unbalanced development of modes • Deficiencies compared to the EU core region • Infrastructure funding alternatives	• Sustainability and innovation • Enlargement and territorial cohesion • Gradual development of main networks (TEN-T, TINA, SEETO)	• Evolving international transport trends
Socio-economic	• Reduced competitiveness of the regional economy • Internal disparities	• Preservation of international competitiveness • Enlargement and convergence	• Financial crisis • Pressures on social welfare
Environmental	• Local pressures • Quality of life	• Congestion • Energy dependency	• Climate change • Availability of energy resources

Source: Own elaboration

6. CONCLUSIVE REMARKS

Throughout the history of regional development, the socio-economic environment interacts with the development of the transportation system. In specific, the development of the transportation system aims at the provision of the accessibility conditions for the service of passenger and freight flows, which are generated by the socio-economic activity of the region. On the other hand, the improvement of accessibility creates new opportunities for the region to develop activities which did not exist or were under-developed prior to this improvement. In this way, transport accessibility is an instrument towards the overcoming of the drawbacks related to the peripheral location of a region. The main precondition is the readiness of the region to capitalise from the improvement of its competitive position in relation to the other regions.

The interaction of the transportation system with regional development was intensified in the era of the international economy and technological innovation. Nowadays, the development of the short and medium distance road transport, the long distance air transport for passenger trips and the intermodality of freight transport networks comprise major contributors for the enhancement of regional competitiveness. At the meantime, the environmental and social pressures imposed by the uncontrolled transport development as well as the impacts from the global economic crisis are leading to the joint efforts of experts and stakeholders for the re-evaluation of priorities and the establishment of new patterns for the sustainable development of the transportation system.

In this changing international environment, the Balkan region is being gradually integrated into the European Union. The European policy framework acknowledges the peripheral position of the Balkans and considers the region's socio-economic convergence as a main goal in order to enhance the international competitiveness of the enlarged European Union. The achievement of this goal depends greatly on the development of the regional transportation system, i.e. the restoration of missing links, the upgrade of infrastructure and services and the reform of the institutional framework, as this is expected to improve the territorial cohesion within the region and throughout the continent.

A decade has passed from the 5th European enlargement and significant progress has been achieved towards the restructuring of the Balkan region's transportation system. However, the statistical data suggests that there is still a significant gap between the Balkans and the rest of the European Union in terms of the availability and quality of infrastructure, the density of the network, the balanced development of modes, the promotion of intermodality, the international competitiveness of major transport terminals and the overall convergence to the principles of sustainable transportation. At the meantime, differentiations are presented in the transport features between different areas within the region, while the uneven spatial distribution of the multimodal transport network indicates the insufficient exploitation of the territorial diversity and the limited accessibility of the region's rural areas. On the other hand, the main European networks, i.e. the trans-European network, the pan-European corridors and the SEETO comprehensive network, provide the conditions of international accessibility in order for the region to build upon its strategic location as the border between three continents and the linkage between the Mediterranean and the Black Seas. The completion of the enlargement process with the accession of all of the Balkan countries to the European Union and the simplification of border-crossing procedures is expected to facilitate the service of flows along the European networks.

Moreover, the preliminary case study for the change in the potential accessibility of road freight transport from the urban centres of Belgrade and Sofia to the international seaport network of the Adriatic and Ionian Seas shows that the expansion of the TEN-T and the accession of the candidate countries is expected to increase the region's accessibility towards its main international gateways, while affecting the attractiveness of specific routes and destinations. An overall conclusion from the case study is that the accession of the Balkan region to the European Union will have various impacts on the regional accessibility of the region by diminishing the time distance of specific routes between major activity poles. In order to be able to monitor these impacts and formulate the corresponding strategies, it is suggested the development and implementation of the appropriate potential accessibility model, with the following features:

- Interdisciplinarity. In the assessment of accessibility, it is required the synthesis of data from various sectors, i.e. transport, socio-economic, spatial, statistical etc. Thus, the formulation of the proposed model should be based on an interdisciplinary approach.

- Specialisation. The model should aim at the monitoring of accessibility in the Balkan region. The selection of its components and the formulation of the implementation scenarios should reflect upon the specific priorities and strategies for the regional development in the Balkans.
- Flexibility. The methodology of the model should be clear and concrete enabling the adjustment to different cases and scopes of analysis.
- Compatibility. The data requirements of the model should be based on official national and European data sources, in order to safeguard the availability of reliable and updated data and to be able to conduct comparisons to the rest of Europe.
- Added value. There is a large inventory of potential accessibility models which have been implemented in various regions of Europe. The proposed model should add value to the existing state of the art by focusing on the peripheral Balkan region.

In this framework, the proposed model can be used as a useful tool for the assessment of the transport features and the related spatial impacts on the Balkan region. In specific, the model can be integrated into:

- The formulation, analysis and evaluation of transport policies.
- The strategic planning of the regional transportation system.
- The dynamic monitoring of the accessibility changes in the context of a regional transport observatory.
- The feasibility analysis from the development or upgrade of transport infrastructure and services.

Finally, the formulation of an integrated potential accessibility model may comprise a common initiative which will enhance the interdisciplinary cooperation of the research community of the Balkan region.

REFERENCES

Albalete D., Bel G. (2010). High-speed rail: Lessons for policy makers from experiences abroad. Working Paper 2010/03, Research Institute of Applied Economics, University of Barcelona (http://www.ub.edu/irea/working_papers/2010/201003.pdf accessed 6/11/2013).

Alfalla-Luque R., Medina-Lopez C. (2009). Supply chain management: Unheard of in the 1970s, core to today's company. *Business History*, 51(2), pp. 202-221.

ARE Transport and Infrastructure Planning and Transport Policy Sections (eds.) (2003). The spatial impacts of transport infrastructures. Learning from the past. Summary of the Project: "Spatial impact of transport infrastructures" initiated by the ARE in collaboration with the Swiss federal agencies concerned and the Swiss cantons. (http://www.are.admin.ch/dokumentation/publikationen/00116/index.html?lang=en&download=NHzLp Zeg7t,lnp6I0NTU042l2Z6ln1ad1IZn4Z2qZpnO2Yuq2Z6gpJCDdIB5gmym1 62epYbg2c_JjKbNoKSn6A-- accessed 14/09/2010).

Baldwin P., Baldwin R. (2004). *The motorway achievement Volume 1. Visualisation of the British motorway system: Policy and administration*. London: Thomas Telford Publishing, London.

Balkan Task Force (2000). *Basic infrastructure investments in South-Eastern Europe*. Regional Project Review. Regional Funding Conference for South-Eastern Europe, European Investment Bank, Brussels.

Baradaran S. (2001). *The Baltic Sea region as a part of Europe. GIS-Analyses of the transport infrastructure and accessibility*. Report: TRITA-IP FR 01-86. Stockholm: Kungl Tekniska Högskolan.

Bartlett W., Prica I. (2012). The variable impact of the global economic crisis in Southeast Europe. LSEE - Research on Southeastern Europe (http://eprints.lse.ac.uk/48037/1/_Libfile_repository_Content_LSEE_Paper s%20on%20South%20Eastern%20Europe_The%20Variable%20Impact(au thor).pdf accessed 12/11/2013).

Brown L. R. (2001). *Paving the Planet: Cars and Crops Competing for Land*. Eco-Economy Update, Washington, DC: Earth Policy Institute.

Bruinsma F., Rietveld P. (1993). Urban agglomerations in European infrastructure networks. *Urban Studies*, 30, pp. 919-934.

Busek E, Kuhne B. (eds.) (2010). *From stabilisation to integration. The Stability Pact for Southeastern Europe.* Bohlau.

Centre for European Reform (2008). *Beyond banking: What the financial crisis means for the EU.* Policy Brief. (http://www.cer.org.uk/sites/default/files/publications/attachments/pdf/2011/pb_fin_crisis_23oct08-787.pdf accessed 29/12/2013).

Committee on Spatial Development of the European Commission (1999). *European Spatial Development Plan (ESDP) towards balanced and sustainable development of the territory of the European Union.* Luxembourg: Office for Official Publications of the European Communities.

Copus A. K. (1999). Peripherality and peripherality indicators. Journal of Nordregio, 10, pp. 11-15.

Copus A. K. (2001). From core-periphery to polycentric development. Concepts of spatial and aspatial periphrality. *European Planning Studies*, 9(4), pp. 539-552.

Dargay J., Gately D., Sommer M. (2007). Vehicle Ownership and Income Growth, Worldwide: 1960-2030. *The Energy Journal*, 28(4), pp. 143-170.

Diego P., (2002). European regional policies in light of recent location theories. *Journal of Economic Geography*, 2(4), Oxford: Oxford Journals, pp. 373-406.

Directorate General for Economic and Financial Affairs (2005). The EU Economy: 2004 Review. 6(2004) (http://ec.europa.eu/economy_finance/publications/publication451_en.pdf accessed 12/11/2004).

DTZ Research (2008). *The rise of the Balkans. European logistics report 2008* (http://www.balkans.com/relevant/c04e7-DTZ_The_Rise_of_the_Balkans_08.pdf accessed 16/11/2013).

ESPON (2003). *Transport services and networks: Territorial trends and basic supply of infrastructure for territorial cohesion.* ESPON Project 1.2.1. 2nd Interim report. (http://www.espon.eu/export/sites/default/Documents/Projects/ESPON2006projects/ThematicProjects/TransportTrends/2.ir_1.2.1-full.pdf accessed 19/12/2013).

ESPON (2004). *Transport services and networks: territorial trends and basic supply of infrastructure for territorial cohesion.* ESPON Project 1.2.1. Project report. (http://www.espon.eu/export/sites/default/Documents /Projects/ESPON2006Projects/ThematicProjects/TransportTrends/fr-1.2.1-full.pdf accessed 29/12/2013).

ESPON (2009). Territorial dynamics in Europe. Trends in accessibility. (http://www.espon.eu/main/Menu_Publications/Menu_TerritorialObservatio ns/trendsinaccessibility.html accessed 8/1/2013).

European Commission (1992). *White Paper. The future development of the common transport policy: A global approach to the construction of a Community framework for sustainable mobility.* COM (92) 494 final, Brussels.

European Commission (1995). *The common transport policy Action Programme 1995-2000.* COM(95) 302 final, Brussels.

European Commission (2001a). *A sustainable Europe for a better world: A European Union strategy for sustainable development.* COM(2001)264 final, Brussels.

European Commission (2001b). *White Paper. European transport policy for 2010: time to decide.* COM(2001) 370 final.

European Commission (2005). *Trans-European transport network. TEN-T priority axes and projects 2005.* Luxemburg: Office for Official Publications of the European Communities.

European Commission (2007a). *Green Paper. Towards a new culture for urban mobility.* COM (2007) 551 final, Brussels.

European Commission (2007b). *Enlargement strategy and main challenges 2007-2008.* COM/2007/0663 final, Brussels.

European Commission (2007c). *Trans-European networks: Towards an integrated approach.* COM(2007) 135 final, Brussels.

European Commission (2007d). *The EU and its neighbouring regions: A renewed approach to transport cooperation.* COM(2011) 415 final, Brussels.

European Commission, (2007e). *Extension of the major trans-European transport axes to the neighbouring countries. Guidelines for transport in Europe and neighbouring regions.* COM(2007) 32 final, Brussels.

European Commission (2008a). *Regions 2020. An assessment of future challenges for EU regions.* Commission Staff Working Document, European Union Regional Policy, SEC(2008). Brussels.

European Commission (2008b). *A European Economic Recovery Plan.* COM(2008) 800 final, Brussels.

European Commission (2008c). *Green Paper on territorial cohesion: Turning territorial diversity into strength.* COM(2008) 616 final, Brussels.

European Commission (2009). *Green Paper. TEN-T: A policy review. Towards a better integrated transEuropean transport network at the service of the common transport policy.* COM(2009) 44 final, Brussels.

European Commission (2010). *EUROPE 2020. A strategy for smart, sustainable and inclusive growth.* COM(2010) 2020 final, Brussels.

European Commission (2011a). *Sustainable development in the European Union. 2011 monitoring report of the EU sustainable development strategy.* Eurostat Statistical Books. Luxembourg: Office for Official Publications of the European Communities.

European Commission (2011b). *White Paper. Roadmap to a single European transport area. Towards a competitive and resource efficient transport system.* COM(2011) 144 final, Brussels.

European Commission (2013a). *Croatia's accession to the European Union - Q&A.* European Commission Memo, 28/06/2013, Brussels (http://europa.eu/rapid/press-release_MEMO-13-629_en.htm accessed 13/11/2013).

European Commission (2013b). *Enlargement strategy and main challenges 2013-2014.* COM(2013) 700 final, Brussels.

European Council (1993). European Council in Copenhagen. Conclusions of the Presidency. Copenhagen, 21-22/06/1993 (http://ue.eu.int/ueDocs/cms_Data/docs/pressdata/en/ec/72921.pdf accessed 24/10/2011).

European Environmental Agency (2011). *Landscape fragmentation in Europe*. Joint EEA-FOEN report. Luxembourg: Publication Office of the European Union.

Eurostat (2009). *Panorama of transport*. Eurostat Statistical Books 2009 edition. Luxembourg: Publications Office of the European Union.

Eurostat (2012). *Energy, transport and environment indicators*. Eurostat Pocketbooks 2012 edition. Luxembourg: Publications Office of the European Union.

Finel N., Tapio P. (2012). *Decoupling transport CO_2 from GDP*. Finland Futures Research Centre FFRC eBOOK 1/2012 (http://www.utu.fi/fi/yksikot/ffrc/julkaisut/e-tutu/Documents/eBook_2012-1.pdf accessed 30/10/2013).

Frank, L. D. (2000). Land use and transportation interaction. Implications on public health and quality of life. *Journal of Planning Education and Research*, 20(1), pp. 6-22.

Gavanas N. (2011). Spatial impacts of the transport system: Implementation for the wider area of South-East Europe. PhD thesis in Greek. Thessaloniki: Aristotle University of Thessaloniki.

Gavanas N., Pavlidou N. (2011). *The impact of the trans-European road network and the process of enlargement on regional accessibility in Southeast Europe*. 51st ERSA Congress: "New Challenges for European Regions and Urban Areas in a Globalised World", 30/8 – 3/9/2011, Barcelona.

Gavanas N., Pitsiava M. (2011). Description of the new member states transport system in an era of convergence – Development of an indicator system. *Spatium International Review*, 24, pp. 37-44.

Gavanas N., Pitsiava-Latinopoulou M. (2013). Methodology for the development of an integrated transport accessibility model for the wider Balkan region. Paper included in the Proceedings of the RESPAG 2013 2nd International Scientific Conference, 22 – 25/5/2013, Belgrade, Serbia (http://iaus.ac.rs/upload/download/Monografije/Conference%20Proceedings-RESPAG.pdf accessed 2/10/2013).

Geneletti D. (2003). Biodiversity Impact Assessment of roads: An approach based on ecosystem rarity. *Environmental Impact Assessment Review*, 23, pp. 343 – 365.

Getimis P., Kafkalas G. (eds.) (2007). *Overcoming fragmentation in Southeast Europe. Spatial development trends and integration potential.* Hampshire: Ashgate. Urban and regional planning and development series.

Gutierrez J., Urbano P. (1996). Accessibility in the European Union: The impact of the trans-European road network, Journal of Transport Geography, 4, pp. 15-25.

Grigor'ev A. N., Severin A. (2007). Debalkanizing the Balkans. A Strategy for a Sustainable Peace in Kosovo. Internationale Politik und Gesellschaft, 1 (http://www.fes.de/ipg/arc_07_set/set_01_07d.htm accessed 27/11/2013).

ICLEI – Local Governments for Sustainability (2011). Strategising sustainable urban mobility in EU neighbour countries. Deliverable of the project: "Increasing energy efficiency of Chişinău and Sevastopol municipalities based on existing positive experience", co-funded by the European Union's CIUDAD Programme (Cooperation in Urban Development and Dialogue) (http://www.iclei-europe.org/fileadmin/templates/iclei-europe/files/content/ Topics/Adaptation/Mobility/Strategising_sustainable_urban_mobility.pdf accessed 14/09/2013).

Informal Ministerial Meeting of Ministers responsible for Spatial Planning and Territorial Development (2011). Territorial agenda of the European Union 2020. Towards an inclusive, smart and sustainable Europe of diverse regions (http://www.eu2011.hu/files/bveu/documents/TA2020.pdf accessed 12/11/2013).

International Transport Forum (2009). *Transport for a global economy. Challenges and opportunities in the downturn.* 2009 Forum Highlights, OECD Publishing (http://www.internationaltransportforum.org/Pub/pdf/ 09Highlights.pdf accessed 20/10/2013).

Jacobsen U., Kristiansen N. (1992). Transport networks for peripheral regions. *Transportation Research Part A: Policy and Practice*, 26(2), pp. 192-210.

Karppi I., Kokkonen M., Lahteenmaki-Smith K. (2001). SWOT-analysis as a basis for regional strategies. Nordregio Working Paper 2001:4 (http://eurolocal.info/sites/default/files/wp0104.pdf accessed 30/11/2013).

Keeble D., Owens P. L., Thompson C. (1982). Regional accessibility and economic potential in the European Community. *Regional Studies*, 16, pp. 419- 432.

Keeble D., Offord J., Walker S. (1988). *Peripheral regions in a Community of twelve Member States.* Luxembourg: Commission of the European Community.

Knowles R., Shaw J., Docherty I. (2008). Transport Geographies: Mobilities, Flows, and Spaces. Oxford: Willey-Blackwell Publishing.

Krugman P. (1998). *Development, geography and economic theory.* Cambridge, Massachusetts and London: MIT Press.

Levinson D., Kanafani A., Gillen D. (1999). Air, high speed rail or highway: A cost comparison in the California corridor. *Transportation Quarterly*, 53 (1), pp. 123-132.

Litman T. (2003). Measuring transportation: Traffic, mobility and accessibility. *ITE Journal*, 73(10), pp. 28-32.

Macario R., Van de Voorde E. (2009). *The impact of the economic crisis on the EU air transport sector.* Brussels: European Parliament.

Maki W. R. (2002). Positioning a metropolitan area for global competition. The Journal of Regional Analysis and Policy, 32(2) (http://jrap-journal.org/pastvolumes/2000/v32/32-2-5.pdf, accessed 24/10/2013).

Nagurney A. (2000). *Sustainable transportation networks.* Northampton: Edward Elgar Pub.

Nemry F., Demirel H. (2012). *Impacts of climate change on transport: A focus on road and rail transport.* JRC Scientific and Policy Reports, Luxembourg: Publications office of the European Union.

Newman P., Kenworthy J. (1999). *Sustainability and cities. Overcoming automobile dependence.* Washington: Island Press.

Nogues S., Salas-Olmedo H. (2009). *Land use and transport integrated policies in peripheral areas*. In Proceeding of European Transport Conference, 5-7/10/2009, Leiden, Netherlands.

OECD (2002). Impact of transport infrastructure investment on regional development. (http://www.internationaltransportforum.org/pub/pdf/02RTRinvestE.pdf accessed 18/10/2012).

Panebianco S. (2001). The impact of European transport infrastructure on peripherality. *Egnatia Odos A.E. Conference on Transport Development and Regional Development*, 5/10/2001, Thessaloniki (http://observatory.egnatia.gr/presentations/PANEBIANCO_2001.pdf accessed 04/09/2011).

Pickton D. W., Wright S. (1998). What's swot in strategic analysis? *Strategic Change*, 7(2), pp. 101–109.

Rodrigue J. P. (2013). *The geography of transport systems (3rd edition)*. New York: Routledge.

Schurmann C., Spiekermann K., Wegener M. (1997). Accessibility Indicators: Deliverable D5 of Project; Socio-economic and spatial Impacts of transport infrastructure investments and transport system improvements (SASI). Commissioned by the General Directorate VII (transport) of the European Commission as part of the 4th Framework Programme of Research and Technology Development. Dortmund: Institut für Raumplanung, Universität Dortmund.

Schurmann C., Talaat A. (2002). *The European peripherality index*. 42nd Congress of the European Regional Science Association (ERSA), 27-31/08/2002, Dortmund.

SEETO Participants (2012). Comprehensive Network development plan (http://www.seetoint.org/wp-content/uploads/downloads/2012/12/SEETO-Comprehensive-Network-Development-Plan-2013.pdf accessed 20/12/2012).

Schurmann C., Talaat A. (2000). *Towards a European Peripherality Index*. Final Report for the General Directorate XVI Regional Policy. Dortmund: Institut fur Raumplanung, Fakultat Raumplanung, Universitat Dortmund.

Spiekermann K., Neubauer J. (2002). European accessibility and peripherality: Concepts, models and indicators (http://www.nordregio.se/en/Publications/Publications-2002/European-Accessibility-and-Peripherality-Concepts-Models-and-Indicators/ accessed 18/10/2012).

Spiekermann K., Wegener M. (2006a). Accessibility and Spatial Development in Europe. *Scienze Regionali*, 5(2) (http://www.spiekermann-wegener.de/pub/pdf/KSMW_Scienze_Regionali.pdf accessed 31/10/2013).

Spiekermann K., Wegener M. (2006b). The role of transport infrastructure for regional development in southeast Europe. *South-East Europe Review for Labour and Social Affairs*, 01/2006, pp. 51-61.

United Nations (2009). *The global economic and financial crisis: Regional impacts, responses and solutions.* New York: United Nations publications.

Venables A., Gasiorek M. (1999). *The welfare implications of transport improvements in the presence of market failure: The incidence of imperfect competition in UK sectors and regions.* Reports to the Standing Advisory Committee on Trunk Road Assessment. London : Dept. of the Environment, Transport and the Regions.

Verny J. (2007). The importance of decoupling between freight transport and economic growth. *European Journal of Transport Infrastructure Research*, 7(2), pp. 113-128.

Vickerman, R. (2004). *Conflicts between transport policies and spatial development policies: perspectives on regional cohesion in the European Union* (http://www-sre.wu-wien.ac.at/ersa/ersaconfs/ersa04/PDF/569.pdf accessed 8/1/2013).

Vickerman, R. (2003). Transport in an integrating Europe: Sustainable development and cohesion. *Investigaciones Regionales*, Seccion Panorama y Debates, 3, pp. 163-174.

Wegener, M. (2008). *SASI model description* (http://www.spiekermann-wegener.de/mod/pdf/AP_0801.pdf accessed 8th January 2013).

Wei S., Yanzi M. (2006). Hub-and-spoke system in air transportation and its implications to regional economic development. A case study of United States. *Chinese Geographical Science*, 16(3), pp. 211-216.

Zhang M., Wiegmans B., Tavasszy L. (2013). Optimization of multimodal networks including environmental costs: A model and findings for transport policy. *Computers in Industry*, 64(2), pp. 136–145.

Printed by Books on Demand GmbH, Norderstedt / Germany